T0329940

Circular Practices in Buildings and Construction to Achieve the Sustainable Development Goals

PROGRESSING THE SUSTAINABLE DEVELOPMENT GOALS SERIES

This timely series offers a multidisciplinary forum for the latest research on critical topics and issues related to the UN's Sustainable Development Goals. Focusing on these global targets and efforts to advance them, books in the series address some of the grand challenges facing society today.

For a full list of Edward Elgar published titles, including the titles in this series, visit our website at www.e-elgar.com.

Circular Practices in Buildings and Construction to Achieve the Sustainable Development Goals

Usha Iyer-Raniga

Professor of Sustainable Built Environment, School of Property, Construction and Project Management, RMIT University, Melbourne, Australia

PROGRESSING THE SUSTAINABLE DEVELOPMENT GOALS SERIES

 Edward Elgar
PUBLISHING

Cheltenham, UK • Northampton, MA, USA

Published by
Edward Elgar Publishing Limited
The Lypiatts
15 Lansdown Road
Cheltenham
Glos GL50 2JA
UK

Edward Elgar Publishing, Inc.
William Pratt House
9 Dewey Court
Northampton
Massachusetts 01060
USA

A catalogue record for this book
is available from the British Library

Library of Congress Control Number: 2024942853

This book is available electronically in the **Elgar**online
Geography, Planning and Tourism subject collection
https://dx.doi.org/10.4337/9781035338870

ISBN 978 1 0353 3886 3 (cased)
ISBN 978 1 0353 3887 0 (eBook)

Printed and bound in Great Britain by
TJ Books Limited, Padstow, Cornwall

Contents

Figures

Tables

Foreword

Finland has been a long-term supporter of multilateral cooperation and partnerships for sustainable development. We have underlined that environmentally, socially and economically sustainable development ultimately depends on the responsible management of the planet's limited natural resources. This is why Finland has vigorously advocated for sustainable consumption and production (SCP) over the years.

The Rio+20 Conference in 2012 reaffirmed that SCP is a cornerstone of sustainable development. One of the very few concrete results from the Rio+20 Conference was the adoption of the 10-Year Framework of Programmes (10YFP) on Sustainable Consumption and Production Patterns (2012–2022). The 10YFP enabled the global community to proceed from vision to action and scale up small individual projects to more effective SCP policies and measures under the United Nations umbrella.

SCP is ultimately a positive agenda. It is not about jeopardizing economic prosperity, but instead about transforming it from a linear to a circular one and doing better with less. SCP is also about fulfilling people's basic needs within the planetary boundaries. The sustainable provision of materials, energy, food, water and shelter is central to ensuring that people are lifted out of poverty and that their wellbeing is maintained or improved.

An increasing number of governments and private companies want to take leadership roles in sustainable and circular solutions. New technologies, however, will not do the job for us if governments do not create opportunities for a fundamental shift to a circular economy and sustainable development.

This book is a story about the journey and results of one thematic programme of the 10YFP, namely Sustainable Buildings and Construction (SBC). The Ministry of the Environment of Finland led the programme from 2015–2022 and ensured its continuation in another format under the Global Alliance on Buildings and Construction (GlobalABC) from 2023 onwards. The work is important because the building sector has enormous potential in reducing global greenhouse gas emissions and material consumption and thereby contributing to sustainable solutions and innovations worldwide. The SBC programme was the first initiative to apply circular economy principles to the global construction and building agenda. These principles have now been included in the work of the Circular Built Environment initiative under the GlobalABC.

Over the years, we have had the pleasure of sharing the leadership together with key experts on sustainable buildings and construction: Professor Usha Iyer-Raniga from RMIT University in Melbourne and Programme Officer Jonathan Duwyn from the United Nations Environment Programme (UNEP). The SBC driving forces in Finland were Senior Architect Harri Hakaste from the Ministry of the Environment and Pekka Huovila, who worked as the determined visionary and Project Coordinator of the SBC programme from the very beginning.

The story of the SBC programme is worth giving thanks to the great commitment and know-how of Pekka, Usha, Jonathan, Harri, members of the SBC Multi-stakeholder Advisory Committee (MAC), the 10YFP Secretariat and many others who contributed to the work.

The legacy of the SBC programme is of great value. I therefore express my gratitude to the authors of this book – Usha Iyer-Raniga and Pekka Huovila – for putting the pieces together. I hope this legacy book will provide new insights and many useful lessons to all readers.

Annika Lindblom

Director for International and EU Affairs
Ministry of the Environment
Finland

Acknowledgements

The intent of this book is to capture the learnings of eight years (2015–2022) of the Sustainable Buildings and Construction (SBC) programme. It has been, and continues to be, an interesting journey, working with a global family of passionate and committed individuals. The SBC programme was set up post the Rio+20 Conference in 2012. This conference reaffirmed that sustainable consumption and production is fundamental to achieve sustainable development. This led to the 10-Year Framework of Programmes (10YFP) on Sustainable Consumption and Production (SCP) (2012–2022). The One Planet Network (OPN) became the implementation platform of the 10FYP, where the SBC programme was but one of six programmes tasked with implementing Sustainable Development Goal 12 of the 2030 Agenda for Sustainable Development.

As with any endeavour, the outputs and outcomes are never a sole pursuit. There are many people to thank. I would like to thank the Government of Finland for taking the lead to drive SBC, and the United Nations Environment Programme (UNEP) for working alongside RMIT University in co-leading the programme. From Finland, I convey my heartfelt thanks to Pekka Huovila, my co-author and coordinator of the SBC programme since its inception. Pekka's unwavering pursuit of delivering on the goals of SBC is to be truly commended. Senior Architect Harri Hakaste from the Ministry of the Environment, Finland, has provided valuable technical support and leadership as we learned and understood the context of sustainability and circularity in the built environment across different regions. Annika Lindblom, Director for International and EU Affairs, Ministry of the Environment Finland, led us through the ups and downs of running a global programme where leads and Co-Leads were strangers before working on SBC collaboratively. She continues to counsel us in the next iteration of SBC. Juulia Suuronen, Specialist at the Unit for International and EU Affairs of the Ministry of the Environment, Finland, and Emma Lappalainen of Finngroup have assisted in supporting various non-technical and technical inputs to broaden and deepen our impacts within the SBC programme, OPN community and beyond.

Being the host of the 10YFP, UNEP has had many people involved in the SBC programme since 2015. My thanks to Programme Officer Jonathan Duwyn, who was involved in the programme from 2017. Fabienne Pierre, Lead Coordinator, One Planet Network Programmes & Flagship Initiatives, 10YFP

Secretariat; and Jorge Laguna Celis, Head of the Secretariat, steered the SBC programme alongside the other OPN programmes. Charles Arden-Clarke and Arab Hoballah were involved in the initial stages of setting up and launching the OPN and creating the OPN family. There are many other members of the UNEP family that have been involved in various activities associated with the OPN programme, such as data collection, communications, marketing, reporting, and so on.

I would like to extend my thanks to the various members of the SBC Multi-stakeholder Advisory Committee, many of whom were part of the programme from the very beginning. Thanks also to the UN intergovernmental organisations, Friends of SCP, members of the 10YFP Board, partners of the SBC and other OPN programmes, and leads and Co-Leads of the OPN programmes; essentially the SBC family and the OPN family that shared a common vision, passion and commitment to support grounded actions to deliver on effective SCP policies at scale.

Finally, I would like to thank RMIT University, who have supported my pro bono commitment to SBC from 2015. I would like to offer my heartfelt thanks to the Dean of Property, Construction and Project Management, Professor Ron Wakefield, for his support in enabling me to work with first the OPN family and now with the GlobalABC family. Last but not least I offer my sincere thanks and gratitude to Priyanka Erasmus for supporting me as Project Officer during the SBC programme.

Usha Iyer-Raniga

Abbreviations

%	percent
10YFP	Ten-Year Framework of Programmes on Sustainable Consumption and Production Patterns
ARBE	Affordable and resilient built environment
ASF	Architecture Sans Frontières
BMZ	Federal Ministry for Economic Cooperation and Development, Germany
C-CAT	Country and Construction Assessment
CBE	Circular Built Environment
CD	Coordination Desk
CE	Circular economy
CI-SCP	Consumer Information Programme for SCP
CO_2	Carbon dioxide
COP	Conference of Parties, UN Climate Change Conference
CRAterre	International Centre on Earthen Architecture, France
CSC	Concrete Sustainability Council
DA	Development Alternatives, India
DRC	Democratic Republic of the Congo
EC	European Commission
ECOSOC	Economic and Social Committee of the UN
eq	Equivalent
EU	European Union
FAO	Food and Agriculture Organization
GCF	Green Climate Fund
GHG	Greenhouse gas
GlobalABC	Global Alliance for Buildings and Construction
GRI	Global Reporting Initiative
GSDR	Global Sustainable Development Report
GSR	Global Status Reports

Habitat III	United Nations Conference on Housing and Sustainable Urban Development
HLPF	High-Level Political Forum on Sustainable Development
IEA	International Energy Agency
IFC	International Finance Corporation
IFI	International Financial Institutions
IKI	International Climate Initiative
IPCC	Intergovernmental Panel on Climate Change
IRP	International Resource Panel
KENSUP	Kenya Slum Upgrading Programme
KFHSD	Khairy Foundation for Human and Social Development
LAC	Latin America and the Caribbean
LCA	Life cycle assessment
LCSA	Life cycle sustainability assessment
MAC	Multi-stakeholder Advisory Committee
MACE	Responsibly sourced materials in circular built environments
MaS-SHIP	Mainstreaming sustainable social housing in India
MoE	Ministry of the Environment, Finland
MPTF	One Planet Multi-partner Trust Fund – financial support to countries implementing Sustainable Development Goal 12
NAP	National Adaptation Plans
NDC	Nationally Determined Contributions
NFP	National Focal Points
NGO	Non-governmental organization
NST	Nabta Smart Town
NZED	Net Zero Energy District
ODA	Overseas Development Aid
OECD	Organisation for Economic Cooperation and Development
OPN	One Planet Network – a global community implementing 10YFP
PUNO	Participating United Nations Organization
PV	Photovoltaics
RE	Resource efficiency
RE-CCAT	Resource Efficient Country and Construction Assessment
RES	Renewable Energy System
SBC	Sustainable Buildings and Construction
SBCI	Sustainable Buildings and Construction Initiative

SBE	Sustainable Built Environment
SCP	Sustainable consumption and production
SDG	Sustainable Development Goal – the 2030 Agenda for Sustainable Development has 17 SDGs
SDG 1	No Poverty
SDG 2	Zero Hunger
SDG 3	Good Health and Well-Being
SDG 4	Quality Education
SDG 5	Gender Equality
SDG 6	Clean Water and Sanitation
SDG 7	Affordable and Clean Energy
SDG 8	Decent Work and Economic Growth – Target 8.4 refers to 10YFP
SDG 9	Industry, Innovation and Infrastructure
SDG 10	Reduced Inequalities
SDG 11	Sustainable Cities and Communities
SDG 12	Responsible Consumption and Production – Target 12.1 is to implement 10YFP
SDG 13	Climate Action
SDG 14	Life Below Water
SDG 15	Life on Land
SDG 16	Peace, Justice and Strong Institutions
SDG 17	Partnerships for the Goals
SDSN	Sustainable Development Solutions Network
SFS	Sustainable Food Systems
SIDS	Small Island Developing States
SLE	Sustainable Lifestyles and Education
SME	Small and medium-sized enterprise
SPP	Sustainable Public Procurement
STP	Sustainable Tourism Programme
t	Tonnes
tCO_2-eq	Tonne carbon dioxide equivalent
TBL	Triple bottom line
TERI	The Energy & Resources Institute, India
UK	United Kingdom
UN	United Nations
UNCT	United Nations Country Team

UNDP	United Nations Development Programme
UNEP	United Nations Environment Programme
UNFCCC	United Nations Framework Convention on Climate Change
UN-Habitat	United Nations Human Settlements Programme
UNOPS	United Nations Office for Project Services
UNWTO	World Tourism Organization – a specialized agency of the United Nations
USD	United States dollar
VAM	Value assessment method
VNR	Voluntary National Reviews
VTT	Technical Research Centre of Finland
WBCSD	World Business Council for Sustainable Development
WCEF	World Circular Economy Forum
WorldGBC	World Green Building Council
WLC	Whole Life Carbon
WMO	World Meteorological Organization
WUF	World Urban Forum – global conference on sustainable urbanization organized by UN-Habitat

1.   Why the SBC programme?

Usha Iyer-Raniga

INTRODUCTION

This first chapter discusses the rationale for why the Sustainable Buildings and Construction (SBC) programme was set up in the first place. The role of buildings and construction has been recognised as a critical contributor to climate change mitigation efforts and, as the limitations of the planet are being acknowledged, in resource efficiency efforts. At the Rio+20 United Nations (UN) Conference in 2012, it was conceded that fundamental changes in our sustainable consumption and production (SCP) patterns are critical to achieve long-term sustainable development. To achieve this global shift to SCP, collaborative and committed diverse actors across the world are needed. This has led to the adoption of the 10-Year Framework of Programmes on Sustainable Consumption and Production Patterns (10YFP). The remit of the 10YFP is to generate collective impact through multi-stakeholder programmes and partnerships, enabling the development, replication and scaling up of SCP policies and initiatives at all levels globally (UNEP 2023a).

The 10YFP consists of six programmes: Sustainable Public Procurement (SPP), Consumer Information (CI) for SCP, Sustainable Tourism Programme (STP), Sustainable Lifestyles and Education (SLE), Sustainable Buildings and Construction (SBC) and Sustainable Food Systems (SFS). The implementation platform of the 10YFP became the One Planet Network (OPN) in 2018. The strategic objective of the OPN is to be recognised as the lead mechanism to support and accelerate the shift to SCP, thus becoming the main operational instrument for Goal 12 of the 2030 Agenda for Sustainable Development (UNEP 2023b). The 10YFP has been documented as part of the strategic objectives and key performance indicators under targets 12.1 and 8.4 of the OPN.

THE URGENCY

Three main interlinked issues that humanity is facing today have been referred to as the triple planetary crisis. These are climate change, pollution and

1

biodiversity loss. The reality facing global citizens today is that our natural resources are in jeopardy, and if current trends continue they will become even more scarce. Biodiversity loss impacts on key basic necessities such as food, water and shelter. Pollution has a direct link to human health and well-being. Climate change is partly due to our increased emissions through fossil-fuel use, and it impacts day-to-day life in the form of changing weather patterns, fluctuations in temperatures, droughts and water scarcity, sea-level rise, storms and declining biodiversity, and impacts our environments through flooding and fires. It also increases our exposure to natural hazards and our ability to respond to these (UNFCCC 2023).

According to UN Secretary-General António Guterres, the era of global warming has ended, and the era of global boiling has begun (UN News 2023). July 2023 has been touted as the warmest month on record. This was confirmed by the World Meteorological Organization (WMO 2023), which said that the planet had had its hottest July, with global ocean surface temperatures hitting a record high for the fourth consecutive month in a row and sea ice being the lowest on record as a result of increased warming.

Anthropogenic emissions are the cause of global warming leading to climate change. Humanity is at a crossroads, facing two issues: curbing greenhouse gas (GHG) emissions and their related problem, climate change. Economic losses have been seen in homes and infrastructure as a result of climate change, and if nothing is done to arrest these, the situation is expected to become more dire (IPCC 2023). In addition to affecting homes and infrastructure, in urban areas climate change has impacted transportation, water, sanitation and energy systems. GHG emissions are a direct result of human activities, largely caused by increasing infrastructure and construction of buildings, focusing attention on cities and their role in the quest for a decarbonised future. In 2015, economic losses were estimated at 314 billion USD globally, and this is expected to rise to 415 billion USD by 2030 (WEF 2022).

In 2019, approximately 79 per cent of global GHG emissions came from a combination of energy, industry, transport and buildings (IPCC 2023). The IPCC also notes that the range of GHG emissions reduction potential is 40–70 per cent across three main sectors: food, land transport and buildings. Up to 66 per cent reduction potential exists in buildings alone (IPCC 2023). However, to reduce GHG emissions impact from buildings, urgent decarbonisation is needed; of prime importance regarding this is efficiency. The move to efficiency in buildings is not going to be cheap, however. Increasing efficiency in buildings may require anywhere from 50–200 USD per tonne carbon dioxide equivalent (tCO_2-eq). Construction material substitutions also bear a cost, but this may be in the range of 20–50 USD per tCO_2-eq, although there is no transparency in data across developed and developing countries (IPCC 2023).

The construction sector accounts for nearly 50 per cent of the total material use across the global economy while also contributing to about 37 per cent emissions (UNEP and Yale Center for Ecosystems + Architecture 2023). Global material use is projected to more than double from 79 Gt in 2011 to 167 Gt in 2060 (OECD 2018). Of the total volume of material used, more than half will be non-metallic materials, while a third is expected to be used in the buildings and construction industry, such as sand, gravel and limestone. There are signs of decoupling between materials use and the economy, as gross domestic product (GDP) will continue to increase but at a much slower rate than the global use of materials. Recycling is projected to become more competitive, establishing a good base for a circular economy as recycling is expected to increase by 3.7 times by 2060 (Iyer-Raniga and Huovila 2020). Also, by 2060, global economic growth in construction and utilities is expected to increase by 2.6 times (OECD 2018). Materials extraction and use has an impact on GHG emissions. By 2060, emissions are expected to reach approximately 50 Gt CO_2-e (OECD 2018). Concrete alone is expected to contribute to 12 per cent of global GHG in 2060 (OECD 2018).

The trends seen to date are not encouraging. Total GHG emissions in the building sector reached 12 $GtCO_2$-eq in 2019, equivalent to 21 per cent of global GHG emissions that year. Over half (57 per cent) of GHG emissions from buildings were indirect CO_2 emissions, arising due to generation of electricity and heat off-site; 24 per cent were direct CO_2 emissions, produced on-site; and 18 per cent were from the production of cement and steel used for construction and refurbishment of buildings. The proportion of emissions related to buildings increased to 31 per cent of global CO_2 emissions. When considering energy alone, the use in residential and non-residential buildings contributed 50 per cent and 32 per cent respectively, while embodied emissions[1] contributed 18 per cent to global building CO_2 emissions (Cabeza et al. 2022).

Over the period 1990–2019, global CO_2 emissions from buildings increased by 50 per cent. At regional levels, emissions in residential buildings decreased in developed countries, except in Australia, Japan and New Zealand, while they increased in developing countries. The highest decreases were observed in Europe and Eurasia, with a 13.6 per cent decrease in direct emissions and 3 per cent decrease in indirect emissions, while the highest increase in direct emissions (of 198 per cent) occurred in the Middle East, and the highest increase in indirect emissions (2258 per cent) occurred in Eastern Asia. Indirect emissions from non-residential buildings increased in all regions. The highest increase, 1202 per cent, occurred in Eastern Asia and the lowest in Europe and Central Asia (4 per cent), where direct emissions from non-residential buildings decreased by 51 per cent (Cabeza et al. 2022).

Embodied emissions have also increased in all regions globally. The highest increase, 334 per cent, occurred in Southern Asia, while the lowest occurred in North America (4 per cent) (Cabeza et al. 2022). Of particular note is the ASEAN (Association of Southeast Asian Nations) region, expected to show very high energy use in the near future. In 2020, energy consumption in buildings accounted for 23 per cent of total final energy consumption in the ASEAN region with 23 per cent of total process and energy-related CO_2 emissions. This is expected to grow by around 60 per cent by 2030 and 120 per cent by 2040, while energy efficiency measures could help to mitigate this growth by at least 20 per cent (IEA 2022a).

Cities may contribute up to 70 per cent of carbon emissions (OECD 2022). Globally, the buildings and construction sector is responsible for 40 per cent of global CO_2 emissions. Of this, about a quarter is associated with building materials production and the remainder is associated with operational energy use over the life of the building (GlobalABC 2022). While both energy and emissions intensity have reduced since 2015, gross floor areas and the cost of investments in buildings went up in the period 2015–2020 (GlobalABC 2022). This is expected to only increase if current trends continue. The Global South in particular will be using a large proportion of building materials in their quest for providing housing and other amenities for a rapidly growing urbanised population. An important consideration is that cities are also facing economic losses as a result of the impacts of climate change.

STATE OF PLAY

In the Global North, due to historical trends and despite energy efficiency regulations in place for several decades, buildings are still not energy-efficient. In the European Union (EU), roughly 75 per cent of the building stock is energy-inefficient (EC 2020). Buildings in the EU are responsible for half of all extracted materials, half of total energy consumption, one-third of water consumption and one-third of waste generation (EC 2022a). So, it is anticipated that in the Global North policy should be directed to focus on deep renovations, encouraging net zero transitions and moving to actual performance of buildings. This fits in well with the ecosystem of decreased populations, mature infrastructure for telecommuting and existing self-sufficient decentralised built environments. The EU Green Deal's focus is on carbon neutrality by 2050 and Fit for 55 (EC 2021), both of which address the whole life carbon (WLC) approach to design, construction, operation and the end of life of buildings.

The Level(s) framework set out by the EC (EC 2022b) is a framework to improve the sustainability of buildings and cities by setting indicators of benchmarks. Level(s) uses core sustainability indicators to measure carbon, materials, water, health, comfort and climate change impacts throughout a building's full life cycle. It is a flexible solution for identifying sustainability hotspots and for future-proofing projects, and assists in policy goals to strengthen the sustainability of Europe's buildings. Level(s) uses free calculation and assessment tools to determine best approaches to design, construction, maintenance and the end of life, and different design and material options, and to make sustainable choices available to consumers (EC 2022c).

There are 16 indicators to measure performance of buildings across three main areas: resource use and environmental performance; health and comfort; and cost, value and risk. This is in line with the New European Bauhaus's sustainability values grounded in the European Commission's new Circular Economy Action Plan (CEAP), adopted in 2020 (EC 2023a). There are three macro-objectives underpinning these sustainability measures: reducing life cycle emissions, improving the resource efficiency and circularity of material life cycles, and optimising the life cycle cost and value of a building, taking into account the market value assessment and risk rating of properties. The circular economy (CE) monitoring framework focuses on material consumption, green public procurement, and waste generation under both production and consumption practices. Under the heading 'Secondary raw materials', recycled materials to meet raw material demand and trading in recyclable raw materials are considered. Under 'Waste management', overall recycling rates and recycling rates for specific streams are included. Under 'Competitiveness and innovation', jobs and value added and innovation are included, and under 'Global sustainability and resilience', consumption footprints and resilience are considered (EC 2023b).

In contrast, in the Global South there is a clear opportunity to leapfrog, to embed circular and low-carbon principles, rather than undergoing the stepped change that the Global North is currently experiencing towards decarbonisation. The tone for construction can be set to highlight not just the environmental considerations but also the social, to support the wider triple bottom line (TBL) considerations of sustainability and directly respond to the Sustainable Development Goals (SDGs).

This leapfrogging tone requires low-carbon building materials and techniques, use of low-emissions materials, upgrading structures and new business models so that long life and durability are promoted, eventually resulting in reductions in the overall carbon footprints. This is where the principles of CE and various strategies associated with optimising outcomes come in. As part of the leapfrogging intent, it is critical to consider holistic approaches so that negative consequences of decisions made are not locked in. For instance, the housing backlog in Africa is at least 51 million units, representing a large potential construction volume in the coming years. At the same time a common issue in many African countries is that building standards and regulation are outdated and insufficient, and their revision represents a great opportunity for including circular principles. When constructing affordable housing the intent should not just be on structural stability but also the materials used so that high-impact emissions are not locked in, and the construction is undertaken in a manner such that these materials may become material banks to be used again (and again) in multiple lives in the future. Such approaches need to be supported by regulation and legislation regarding materials and building codes.

The United Nations Environment Programme (UNEP 2022) states that due to the increase in indirect emissions, four important areas need to be focused on going forward: minimise excess floor area, reduce the energy intensity of buildings, reduce emissions intensity and reduce embodied emissions from construction. Along similar lines, the Global Alliance for Buildings and Construction (GlobalABC), International Energy Agency (IEA) and UNEP (2020a) outlined a list of actions in the short, medium, and long term to make the built environment zero-emission, efficient and resilient. The way forward includes actions, targets, policies, technologies and enabling measures across urban planning, new buildings, existing buildings, building operations, appliances and systems, materials, resilience and clean energy. For urban planning, alignment across different scales and prioritising energy efficiency is critical. For new buildings, mandatory energy efficiency codes, high-performance buildings and high efficiency standards need to be developed, implemented and enforced.

For existing buildings, deep renovations and retrofits that increase decarbonisation in buildings are sought. For building operations, adoption of energy performance tools, improved operations, and continuous monitoring, evaluation and feedback are critical. For appliances and systems, prioritising energy efficiency in public procurement and stimulating demand for energy-efficient appliances need to be considered. Materials that are low-carbon and promote efficiency, reduced embodied energy and whole life cycle considerations need attention. For resilience, adaptations of new and existing buildings need to consider integrating resilience, and for clean energy encouraging on-site renewal, procurement of green power, and moving to renewables to operate buildings and for cooking are the focus. These areas of concern are also highlighted in various roadmaps developed across the Arabic region (Bhar et al. 2023), Africa (GlobalABC/IEA/UNEP 2020b), Asia (GlobalABC/IEA/UNEP 2020c) and Latin America (GlobalABC/IEA/UNEP 2020d).

The measures for decarbonising are in sync with expectations across some countries. Key findings of an Organisation for Economic Co-operation and Development survey (OECD 2022) included the importance of decarbonising for creating jobs, enhancing well-being and increasing energy affordability at local levels. While cities and regions have their own plans for decarbonisation, implementation was identified as a critical issue. Poor monitoring, poor evaluation frameworks and fragmentation result in obstacles to turning strategies into action. Funding is a problem and there are capacity gaps, wherein small and medium-sized enterprises (SMEs) find it difficult to get access to information and financing. Overall, 88 per cent of the cities and regions surveyed for the OECD report wanted to implement higher energy efficiency standards than the national level in building energy codes, with a quarter calling for net zero energy levels. However, the enforcement of mandatory building energy codes and developing effective regulations for existing buildings were also highlighted. Collaboration across various levels of government was cited as being very important for a critical role in enabling policy environments for cities and regions.

In 2019, the IEA (2019) estimated that a clean energy future is needed to increase investment in buildings. If timely investment is made in buildings, it will ultimately bring savings across the global economy and cut 50 per cent of household income spent on energy. But to do so, annual capital flows are required. The estimated budget for new buildings plus changes to existing buildings was projected to increase by an average of 27 billion USD over the 2020–2030 period, a relatively small addition to the 4.9 trillion USD already invested each year in buildings globally. Cumulative household energy spending to 2050 can be around 5 trillion USD lower, leading to net savings for consumers, with the average share of household income spent on energy (2019 figures) falling from 5 per cent to around 2.5 per cent by 2050 (IEA 2019).

The IEA estimates that achieving net zero by 2050 requires the buildings sector to realise an average energy intensity improvement of 4–5 per cent per year between 2020 and 2050. By 2025, buildings should move to all-electric; by 2030, all new buildings should be zero-carbon-ready. By 2040, all existing buildings should be retrofitted to achieve zero-carbon-ready levels by 2050. More than 85 per cent of buildings are zero-carbon-ready (IEA 2021, IEA 2022b) in theory but still need to demonstrate this through performance. The *World Energy Outlook* (IEA 2022b) also highlighted the importance of policy response as a part of the approach to achieve the 1.5 degrees C temperature rise under the Paris Agreement.

In 2022 the World Green Building Council (WorldGBC) presented its *EU Policy Whole Life Carbon Roadmap*, where it considered four main areas for achieving whole of life carbon: building regulations, waste and circularity, sustainable procurement and sustainable finance. Under 'Building regulations', minimum performance standards and energy performance certificates alongside the setting up of databases for tracking and monitoring, and whole-of-life and energy-performance-of-buildings directives, were considered. Under 'Waste and circularity', energy-performance-of-buildings directives, construction products regulation, waste frameworks, essential tools for circularity and voluntary sectoral standards were the main factors considered. 'Sustainable procurement' considered guidance, tender criteria, EU procurement directives, and national and regional policies. 'Sustainable finance' included taxonomy and renovation finance.

CONTEXT FOR THE SBC PROGRAMME

It is against this background that the SBC programme was situated. The programme was launched in April 2015, before the SDGs came into existence. It was set up with the vision to achieve a situation in 2030 where:

> All stakeholders in the planning, design, construction, commissioning, management, operation and deconstruction of buildings have a common understanding of sustainable buildings and the knowledge, resources and incentives required to create, maintain and use them; structures that are healthy to live and work in, that sustainably utilise energy, water, and other key resources, respecting environmental limits, and ultimately have a minimally adverse impact on the natural world, supporting social and economic development. (SBC programme website)

The next chapter dives deeper into the SBC programme, and its background, aims and objectives.

NOTE

1.   Embodied emissions comprise emissions that are used to make the materials that are used for building. Typically, this covers scope 1 and scope 2 emissions. Finding ways to include scope 3 emissions is still a matter of debate.

REFERENCES

Bhar, R., Mahdy, E., and Hermelink, A. (2023) *GlobalABC Regional Roadmap for Buildings and Construction in the Arab Region 2020–2050 Towards a Zero-Emission, Efficient, and Resilient Buildings and Construction Sector*, Ministry of Energy and Infrastructure, Guidehouse and RCREEE. https://globalabc.org/sites/default/files/inline-files/GlobalABC_Roadmap_for_Buildings_and_Construction_2020-2050_3.pdf

Cabeza, L.F., Bai, Q., Bertoldi, P., Kihila, J.M., Lucena, A.F.P., Mata, E.M., Irasgedis, S., Novikova, A., and Saheb, Y. (2022) Buildings. In IPCC, *Climate Change 2022: Mitigation of Climate Change – Contribution of Working Group III to the Sixth Assessment Report of the Intergovernmental Panel on Climate Change* [P.R. Shukla, J. Skea, R. Slade, A. Al Khourdajie, R. van Diemen, D. McCollum, M. Pathak, S. Some, P. Vyas, R. Fradera, M. Belkacemi, A. Hasija, G. Lisboa, S. Luz and J. Malley, (eds.)], Cambridge University Press, Cambridge, UK and New York, NY. https://doi.org/10.1017/9781009157926.011

EC (European Commission) (2020) *Energy Efficiency in Buildings*, Publications Office of the European Union, Luxembourg. https://commission.europa.eu/system/files/2020-03/in_focus_energy_efficiency_in_buildings_en.pdf

EC (European Commission) (2021) *Communication from the Commission to the European Parliament, The Council, The European Economic and Social Committee and the Committee of Regions, 'Fit for 55' Delivering the EU's 2030 Climate Target on the Way to Climate Neutrality*, Publications Office of the European Union, Luxembourg. https://eur-lex.europa.eu/legal-content/EN/TXT/PDF/?uri=CELEX: 52021DC0550

EC (European Commission) (2023b) *Circular Economy Action Plan*, Publications Office of the European Union, Luxembourg. https:// environment .ec .europa .eu/ strategy/circular-economy-action-plan_en

EC (European Commission) Directorate-General for Environment (2022a) *Level(s): European Framework for Sustainable Buildings*, Publications Office of the European Union, Luxembourg. https:// environment .ec .europa .eu/ topics/ circular -economy/ levels_en

EC (European Commission) Directorate-General for Environment (2022b) *Level(s): What's in It for Cities?*, Publications Office of the European Union, Luxembourg. https://data.europa.eu/doi/10.2779/949982

EC (European Commission) Directorate-General for Environment (2022c) *The Level(s) Calculation and Assessment Tool (CAT)*, Publications Office of the European Union, Luxembourg. https://data.europa.eu/doi/10.2779/109620

EC (European Commission) Directorate-General for Environment (2023a) *Circular Economy: New Tool for Measuring Progress*, Publications Office of the European Union, Luxembourg. https://data.europa.eu/doi/10.2779/1618

GlobalABC (Global Alliance for Buildings and Construction) (2022) The Global Alliance for Buildings and Construction: An Overview Over Status and Achievements. https://globalabc.org/

GlobalABC/IEA/UNEP (Global Alliance for Buildings and Construction, International Energy Agency and United Nations Environment Programme) (2020a) *GlobalABC Roadmap for Buildings and Construction 2020–2050: Towards a Zero-Emission, Efficient and Resilient Buildings and Construction Sector*. https:// globalabc .org/ sites/ default/ files/ inline -files/ GlobalABC _Roadmap _for _Buildings _and _Construction_2020-2050_3.pdf

GlobalABC/IEA/UNEP (Global Alliance for Buildings and Construction, International Energy Agency and United Nations Environment Programme) (2020b) *GlobalABC Regional Roadmap for Buildings and Construction in Africa 2020–2050: Towards a Zero-Emission, Efficient and Resilient Buildings and Construction Sector*. https:// globalabc.org/sites/default/files/inline-files/Africa_Buildings%20Roadmap_FINAL _1.pdf

GlobalABC/IEA/UNEP (Global Alliance for Buildings and Construction, International Energy Agency and United Nations Environment Programme) (2020c) *GlobalABC Regional Roadmap for Buildings and Construction in Asia 2020–2050: Towards a Zero-Emission, Efficient and Resilient Buildings and Construction Sector*. https:// globalabc.org/ sites/ default/ files/ inline-files/ Asia_Buildings%20Roadmap_FINAL .pdf

GlobalABC/IEA/UNEP (Global Alliance for Buildings and Construction, International Energy Agency, and the United Nations Environment Programme) (2020d) *GlobalABC Regional Roadmap for Buildings and Construction in Latin America 2020–2050: Towards a Zero-Emission, Efficient and Resilient Buildings and Construction Sector*. https:// globalabc .org/ sites/ default/ files/ inline -files/ 2 . %20GlobalABC_Regional_Roadmap_for_Buildings_and_Construction_in_Latin _America_2020-2050.pdf

IEA (International Energy Agency) (2019) *Perspectives for the Clean Energy Transition: The Critical Role of Buildings.* https://iea.blob.core.windows.net/assets/ 026bff1b -821d -48bc -8a0e -7c10280c62bc/ Perspectives _for _the _Clean _Energy _Transition_2019.pdf

IEA (International Energy Agency) (2021) *Net Zero by 2050: A Roadmap for the Global Energy Sector.* www.iea.org/reports/net-zero-by-2050

IEA (International Energy Agency) (2022a) *Roadmap for Energy-Efficient Buildings and Construction in the Association of Southeast Asian Nations.* https:// www .iea .org/ reports/ roadmap -for -energy -efficient -buildings -and -construction -in -the -association-of-southeast-asian-nations

IEA (International Energy Agency) (2022b) *World Energy Outlook 2022,* revised version. https://www.iea.org/reports/world-energy-outlook-2022

IPCC (Intergovernmental Panel on Climate Change) (2023) Summary for Policymakers. In *Climate Change 2023: Synthesis Report – Contribution of Working Groups I, II and III to the Sixth Assessment Report of the Intergovernmental Panel on Climate Change.* https://doi.org/10.59327/IPCC/AR6-9789291691647.001

Iyer-Raniga, U., and Huovila, P. (2020) *Global State of Play for Circular Built Environment: A Report on the State of Play on Circularity in the Built Environment across Africa, Asia, Europe, Gulf Cooperation Council Countries, Latin America and the Caribbean, North America and Oceania – Final Report October 2020,* United Nations One Planet Network Sustainable Buildings and Construction Programme. https://www.oneplanetnetwork.org/knowledge-centre/resources/global -state-play-circular-built-environment

OECD (Organisation for Economic Co-operation and Development) (2018) *Global Material Resources Outlook to 2060: Economic Drivers and Environmental Consequences – Highlights.* https:// www .oecd .org/ env/ global -material -resources -outlook-to-2060-9789264307452-en.htm

OECD (Organisation for Economic Co-operation and Development) (2022) *Decarbonising Buildings in Cities and Regions,* OECD Urban Studies. https:// doi .org/10.1787/a48ce566-en

UNEP (United Nations Environment Programme) (2022) *Emissions Gap Report 2022: The Closing Window – Climate Crisis Calls for Rapid Transformation of Societies.* https://www.unep.org/emissions-gap-report-2022

UNEP (United Nations Environment Programme) (2023a) 10YFP: 10-Year Framework of Programmes on Sustainable Consumption and Production Patterns. https://www .unep.org/explore-topics/resource-efficiency/what-we-do/one-planet-network/10yfp -10-year-framework-programmes

UNEP (United Nations Environment Programme) (2023b) One Planet Network. https:// www.unep.org/explore-topics/resource-efficiency/what-we-do/one-planet-network

UNEP (United Nations Environment Programme) and Yale Center for Ecosystems + Architecture (2023) *Building Materials and the Climate: Constructing a New Future.* https://wedocs.unep.org/20.500.11822/43293

UNFCCC (United Nations Framework Convention on Climate Change) (2023) What Is the Triple Planetary Crisis? https:// unfccc .int/ news/ what -is -the -triple -planetary -crisis

UN News (2023) Hottest July Ever Signals 'Era of Global Boiling Has Arrived' Says UN Chief. https://news.un.org/en/story/2023/07/1139162

WEF (World Economic Forum) (2022) *Delivering Climate-Resilient Cities Using a Systems Approach,* Future of Cities Reports 1/4, Global Future Council on Cities of

Tomorrow. https://www3.weforum.org/docs/WEF_C4IR_GFC_on_Cities_Climate _Resilience_2022.pdf

WorldGBC (World Green Building Council) (2022) *EU Policy Whole Life Carbon Roadmap for Buildings*. https:// worldgbc .org/ article/ eu -policy -whole -life -carbon -roadmap-for-buildings/

WMO (World Meteorological Organization) (2023) July 2023 Confirmed as Hottest Month on Record. https:// public .wmo .int/ en/ media/ news/ july -2023 -confirmed -hottest-month-record

2. History and background of the SBC programme

Usha Iyer-Raniga

INTRODUCTION

This chapter provides the history and background of the SBC programme. The previous chapter provided the rationale for why the programme was set up, and the background to the OPN. The OPN harnesses expertise across high-impact systems and sectors to help provide the enabling conditions for a transformation of how we consume and produce. The network is a global community of practitioners, policymakers and experts, including governments, businesses, civil society, academia and international organisations, that implements the 10YFP on SCP and works towards achieving SDG 12: ensuring sustainable patterns of consumption and production.

AIM OF THE SBC PROGRAMME

The SBC programme was aimed at improving the knowledge of sustainable construction and supporting and mainstreaming sustainable building solutions. Through the programme, all major sustainable construction activities were brought together under the same umbrella. The work involved sharing good practices, launching implementation projects, creating cooperation networks and committing actors around the world to sustainable construction.

The programme was underpinned by guiding principles, recognising the urgency of SBC. These principles enumerated, in no particular order (SBC 2015):

1. Promoting common language and tools on 'sustainable building'. SBC is more than green or energy-efficient buildings. While many policies aim to reduce energy use in buildings, a key component of SCP, buildings should contribute more broadly to the three pillars of sustainability or TBL (environmental, social and economic), including resource efficiency. A first step is for stakeholders to reach consensus on what is a 'sustainable

building', which can be achieved through increased dialogue and the development of a 'common language' allowing for broader uptake of SBC.

2. Adopting a life cycle approach and resource efficiency in building supply chains. The challenge is to make life cycle assessment (LCA) more efficient and effective, with effectual rules and guidance allowing for broader uptake, especially in developing nations. Broader adoption of LCA and resource efficiency shapes the planning and design of buildings by enabling the building supply chain to improve its performance through the greater inclusion of renewable resources and energy, and water-efficient approaches, and by reducing the non-renewable use of natural resources.

3. SBC should use scientific and technological knowledge to produce tools that are designed for field actors and lead to cost-effective policies. Research methodologies exist regarding resource efficiency, resilience to climate change and other challenges. Yet, field actors such as developers, architects, engineers and contractors are looking for user-friendly, affordable and cost-effective approaches. In a context of economic crises and a lack of public resources in many parts of the world, affordable and cost-effective actions are the best way to spread the SBC approach.

4. SBC depends on sustainable infrastructure. Buildings should be designed and constructed to respond to available infrastructure and local environmental and socio-economic conditions, and strive to drive demand and enable the provision of more sustainable infrastructure and urban services.

5. SBC should strive to be near-zero-impact, or even, when relevant, 'resource positive'. Buildings should aim to use near-zero energy and water and to produce near-zero waste. When this is relevant, and taking into account their urban context and other factors, they can even provide more energy than they consume; they should aim to produce more clean water than they use and aim to utilise more waste than they produce. While technologies exist today to make this happen, from 'living machines' that treat wastewater to the use of agricultural waste products in building materials and fuel stocks, there remains more to be done in research, pilot applications and enabling policies.

6. SBC should have a circular economy approach. The economic case for energy- and resource-efficient buildings has been made and is increasingly understood. Yet more needs to be done to develop and promote the case for sustainable buildings as part of the circular economy, and the transition towards more sustainable buildings and construction will necessitate changes from the way building projects are contracted and financed to how organisations budget and governments incentivise the right kinds of actions at the right times throughout the full life cycle of a building.

7. SBC should begin at home. To date, most 'green-certified buildings' have involved commercial, institutional and public sector buildings. However, more than 1.8 billion people worldwide already lack adequate housing and every year a further 2 million people are forcibly evicted, and some 150 million people worldwide are homeless (UN-Habitat 2023). Further, as a majority of our time is spent at home a strong focus on housing will allow greater positive impact and a greater increase in the sustainability of our lifestyles.

8. SBC requires engaged and knowledgeable active participation of all stakeholders. The success of a building's performance depends on its proper operation. Unfortunately, a number of 'green-labelled' buildings suffer from a gap between the designed use and actual performance. Building users and facility managers are one of the most critical components for ensuring the success of a sustainable building. There is a need to increase stakeholders' knowledge, access to consumer and market information, training, and understanding of the important environmental, social and economic opportunities in the building sector to ensure a building performance focus that avoids 'locking in' poor performance.

9. SBC necessitates a climate-responsive approach. Buildings last longer than almost all other types of products, and the world around them is expected to undergo great change during their lifetime. They should be designed to help their occupants cope with observed and anticipated climate change impacts (adaptation), and efforts to further reduce their impact must be supported (mitigation).

10. Monitoring, verifying and providing feedback is critical for success. The sustainability performance of buildings and infrastructure throughout their life cycle needs to be verified through monitoring and providing feedback to ensure near- and long-term goals are accomplished. Additional tools, methods and approaches are needed to ensure this can occur during all stages of the building's life.

11. There is a key role for government at all levels to 'lead by example'. Amongst public building inventories such as schools, hospitals and other public stock, as well as financially subsidised programmes (e.g. affordable housing), there should be exemplars of sustainable building and catalytic agents to help develop resource-efficient supply chains, technologies, skills and expertise in the construction sector and to educate and inform their occupants about sustainability and further the development of a multi-stakeholder approach.

12. Demonstration of SBC approaches is essential. Pilot projects and other 'hands-on' demonstrations provide encouragement to stakeholders and valuable lessons learned, and help to refine approaches. Demonstration projects should not only focus on the environmental improvements but

also identify and consider economic effects (job creation), financial impacts (costs and savings) and other social aspects (work health and safety). Further, case studies and reports should be encouraged to provide a 'warts and all' detailing of lessons learned so that successes can be replicated, and obstacles identified and avoided.

The SBC programme was formally approved by the 10YFP Board at the commencement of the programme in 2015. The objective of the programme was to support an enabling environment for SBC by strengthening the SCP knowledge base and methodologies, engaging relevant stakeholders, and enabling the definition and implementation of supportive policies and regulations. Four different work streams were identified:

WS1: Establish and promote enabling frameworks to implement SBC policies
The following were needed to achieve this:

- Fostering and sharing research, tools and financial and other approaches related to SBC
- Maintaining and engaging in global dialogue to develop and promote common language and tools related to SBC
- Fostering enabling frameworks for SBC

WS2: Support and promote sustainable housing
The main goals for sustainable housing were:

- Piloting sustainable housing approaches in the affordable and social housing markets
- Promoting SBC policies for the low-income housing sector
- Supporting synergies with relevant programmes

WS3: Enhancing sustainability in the building supply chain
Key factors considered were:

- Identifying and sharing core analytical tools, conceptual work and improved knowledge bases for supporting decision making towards resource efficiency in building supply chains
- Promoting policies to integrate resource efficiency in building supply chains

WS4: Reducing climate impact and strengthening climate resilience of the buildings and construction sector
The main considerations under this work stream were:

• Identifying and sharing core analytical tools, conceptual work and improved knowledge bases for supporting decision making towards a more climate-resilient and low-emission buildings and construction sector
• Promoting resilient and low-emission SBC planning and piloting where appropriate.

In addition to these work streams, a cross-cutting theme was identified. This focused on promoting knowledge sharing, outreach and awareness raising. The objectives of this cross-cutting work stream were to increase awareness of the SBC programme and its focus areas and results, and to provide a knowledge-sharing platform to exchange lessons and experiences amongst relevant stakeholders and also the general public. This was done using various events, green-building conferences, workshops and seminars organised by educational institutes as well as lifelong learning programmes.

This cross-cutting stream consisted of four themes:

1. Promoting awareness-raising efforts and understanding of sustainable buildings and construction across the stakeholders and general public
2. Supporting the organisation of international conferences and platforms for disseminating activities and results
3. Supporting the building of peer groups based on region-specific conditions and sharing lessons and experiences with building codes, solutions and geographically bound infrastructure
4. Promoting interdisciplinary exchanges between institutions of higher education for architecture, urban planning, engineering and the like, and informing the development and application of SBC policies while fostering integration of sustainability considerations in relevant curricula supported by concrete case studies.

Having discussed the objectives and work streams, the next section discusses the SBC programme's structure.

PROGRAMME STRUCTURE

At the start of the SBC programme, the Ministry of the Environment (MoE) of Finland led it, supported by three Co-Leads: UNEP, RMIT University and the World Green Building Council (WorldGBC)[1] (see Figure 2.1). The Coordination Desk (CD) was comprised of the Leads and Co-Leads of the SBC programme.

10YFP SBC PROGRAMME

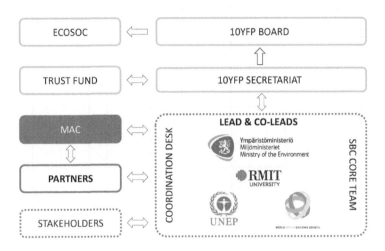

Figure 2.1   *Key stakeholders and their roles in the SBC programme in 2015*

The programme was supported by Multi-stakeholder Advisory Committee members (MACs), as listed in Table 2.1, ensuring global coverage and a range of organisational types and scales. Organisations were operating at national, regional or international scales or across government, business, academia or inter-governmental organisation (IGO) levels or as a non-governmental organisation (NGO). A total of 23 organisations, including the Lead/Co-Leads, comprised the SBC programme in 2015.

From the start, the programme recognised the challenges regarding different time zones, unequal capabilities and resources, a lack of appropriate technologies and different working cultures and expectations. It was agreed from the start that meetings would take place online, using mobile technologies.

During the period 2015–2022, the number of MACs and partners varied. However, a core group of MACs and partners remained committed to the SBC programme. Workplans were undertaken on an annual or a bi-annual basis to steer the programme and ensure deliverables were met.

Table 2.1 SBC programme members at the commencement of the programme

	Role	Organization	Government	Business	Academia	IGO	NGO	National	Regional	International	Asia	Africa	LAC*	Europe	East Europe	North America	Oceania
1	Lead	MoE, Finland	1					1						1			
2	Co-Lead	RMIT, Australia			1			1									1
3	Co-Lead	UNEP, France				1				1				1			
4	Co-Lead	WorldGBC, US					1			1						1	
5	MAC	Building and Construction Authority (BCA), Singapore	1					1			1						
6	MAC	The Energy & Resources Institute (TERI), India					1			1	1						
7	MAC	Jabatan Kerja Raya Malaysia (Malaysian Department of Public Works)	1					1			1						
8	MAC	Bioregional, UK					1			1				1			
9	MAC	Energies 2050 partnered with tiers de l'Architecture et d'Urbanisme (EAMAU), France			1				1					1			
10	MAC	French Ministry of Ecology	1					1						1			
11	MAC	Polito, Italy			1			1						1			

#		Organisation	Oceania	North America	East Europe	Europe	LAC*	Africa	Asia	International	Regional	National	NGO	IGO	Academia	Business	Government
12	MAC	World Business Council for Sustainable Development (WBCSD), Switzerland				1				1						1	
13	MAC	UNEP-SBCI (Sustainable Buildings and Construction Initiative), France				1				1			1				
14	MAC	UN Office for Project Services (UNOPS), Kenya						1		1				1			
15	MAC	Construction Industry Development Board (CIDB), South Africa						1				1					1
16	MAC	Skanska, Sweden				1				1						1	
17	MAC	South East Energy Efficiency Alliance (SEEA), US		1							1		1				
18	MAC	UN-Habitat, Kenya						1		1				1			
19	MAC	Development Alternatives (DA), India							1	1			1				
20	MAC	RICS, Belgium				1				1			1				
21	MAC	World Wildlife Fund (WWF), Switzerland				1				1			1				

			Government	Business	Academia	IGO	NGO	National	Regional	International	Asia	Africa	LAC*	Europe	East Europe	North America	Oceania
22	MAC	Tsinghua University, China	1		1			1			1						
23	MAC	Ministry of Production and Labor, Argentina						1					1				
Total			6	2	4	3	8	9	2	12	5	3	1	11	0	2	1

Note: * Latin America and the Caribbean.

SBC WORKPLANS

The programme commenced with shared responsibility across the SBC leadership team. Both geographical spread and activities were shared as shown in Figure 2.2.

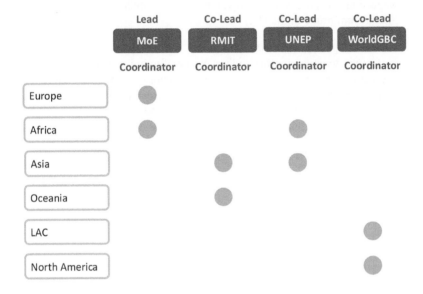

Figure 2.2   Geographical spread and Coordination Desk roles in the SBC programme in 2015

In addition, the Secretariat also supported the SBC programme with the following key functions:

- Providing a platform for online questionnaires, and supporting identification of relevant and interested donors, ensuring regional and stakeholder balance
- Providing platforms and opportunities for the programmes in convening working groups and training to meet relevant demand and needs
- Identifying synergies between the 10YFP programmes and the existing regional SCP strategies
- Coordinating calls for proposals under the Trust Fund of the 10YFP and providing support for fundraising activities
- Providing the Global SCP Clearinghouse, as a dynamic and decentralised portal

- Inviting stakeholders to contribute to relevant reports and publications, where they could showcase their best practices and work
- Convening meetings and webinars with programme Leads/Co-Leads, MACs and partners to share experience and best practices and explore areas of cooperation
- Reporting the progress of programmes to ECOSOC (the Economic and Social Committee of the UN).

The first workplan covered the period 2015–2016. The main focus of this work plan was to develop project ideas, explore fundraising opportunities, and engage with the programme and partners to support collaborative approaches across the various work streams (see Figure 2.3). Specific events and outreach activities associated with SBC were also highlighted.

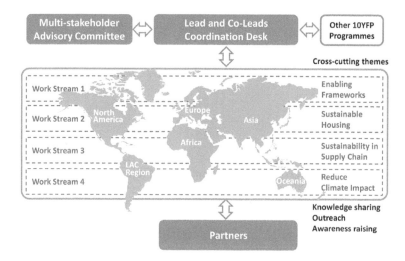

Figure 2.3   SBC work streams and networks

The 2017–2018 workplan focused on forming closer links with the financing community to support fundraising; achieving impacts through the projects in terms of outcome, replicability and scalability for the Trust Fund projects; engaging in resource mobilisation to commence flagship projects; and making progress with other SBC projects. The workplan also outlined outreach and promotion activities, communicating progress on projects and collaboration with the other programmes across the OPN and the Global Alliance on Buildings and Construction (GlobalABC).

Communication, in particular, was organised across the following main themes:

- Sustainable built environment is the best investment option for climate change mitigation and adaptation.
- The 10YFP SBC is the best platform to coordinate and implement sustainable construction globally.
- The SBC programme is the go-to place for good sustainable building practices.
- The SBC programme supports ongoing sustainable activities and develops new projects.
- The SBC programme provides information about funding sources and clients and partners in the network.
- The SBC programme wants to try new technologies in a collaborative/partnership arrangement.
- The SBC programme monitors impacts using a common 10YFP SCP assessment framework and metrics.

The 10YFP Board supported the programmes with funds, including the SBC programme. The SBC 10YFP Trust Fund call for proposals was announced in 2015 at COP 21, with high visibility. By February 2016, 108 proposals were received, and the best proposals were submitted to the 10YFP Board for approval in April 2016. By June of that year, the three chosen proposals moved forward to contract negotiations. Additional funding was provided in 2017 for two more proposals. Thus, in total five projects were seeded by the SBC programme in 2017. Most of these projects were implemented over a period of 18–24 months. More information on the projects is provided in the next chapter.

In 2019, the SBC programme was expanded from housing/buildings to the built environment. Two thematic areas were identified to work on to the end of the programme. These were:

1. Affordable and resilient built environment (ARBE): Under this thematic focus area, the aim of the SBC programme was to advocate for and promote affordable, sustainable and resilient housing construction practices, with a focus on holistic and integrated urban and community development.
2. Responsibly sourced materials in circular built environments (MACE): Under this thematic focus area, the SBC programme's objective was to promote circularity principles and approaches in the buildings and construction sector to support countries in achieving their climate commitments, environmental commitments and development objectives, notably through the promotion of responsibly sourced materials and the reuse of different waste streams.

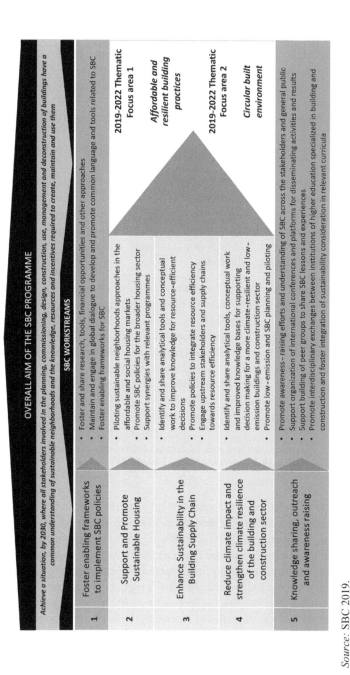

OVERALL AIM OF THE SBC PROGRAMME

Achieve a situation, by 2030, where all stakeholders involved in the planning, commissioning, design, construction, use, management and deconstruction of buildings have a common understanding of sustainable neighborhoods and the knowledge, resources and incentives required to create, maintain and use them

SBC WORKSTREAMS

1	Foster enabling frameworks to implement SBC policies	• Foster and share research, tools, financial opportunities and other approaches • Maintain and engage in global dialogue to develop and promote common language and tools related to SBC • Foster enabling frameworks for SBC
2	Support and Promote Sustainable Housing	• Piloting sustainable neighborhoods approaches in the affordable and social housing markets • Promote SBC policies for the broader housing sector • Support synergies with relevant programmes
3	Enhance Sustainability in the Building Supply Chain	• Identify and share analytical tools and conceptual work to improve knowledge for resource-efficient decisions • Promote policies to integrate resource efficiency • Engage upstream stakeholders and supply chains towards resource efficiency
4	Reduce climate impact and strengthen climate resilience of the building and construction sector	• Identify and share analytical tools, conceptual work and improved knowledge bases for supporting decision making for a more climate-resilient and low-emission buildings and construction sector • Promote low–emission and SBC planning and piloting
5	Knowledge sharing, outreach and awareness raising	• Promote awareness–raising efforts and understanding of SBC across the stakeholders and general public • Support organization of international conferences and platforms for disseminating activities and results • Support building of peer groups to share SBC lessons and experiences • Promote interdisciplinary exchanges between institutions of higher education specialized in building and construction and foster integration of sustainability consideration in relevant curricula

2019-2022 Thematic Focus area 1

Affordable and resilient building practices

2019-2022 Thematic Focus area 2

Circular built environment

Source: SBC 2019.

Figure 2.4   The five work streams of the SBC programme

Each thematic focus area had a dedicated working group and a set of commonly agreed activities and targets. Through its work in each of its thematic focus areas the programme sought to (i) make the case for SBC to countries, (ii) provide training and technical and policy guidance based on science to key stakeholders for shifting policies and actions towards the promotion of sustainable buildings and construction, and (iii) facilitate the creation partnerships for SBC implementation. These are explained briefly in Figure 2.4.

These activities continued alongside the development of the state-of-play reports on circular built environments that were launched in 2020, and deep dives in the Asian, African and Latin American regions continued into 2021–2022 (SBC 2023).

To build commitment and foster global action for making 'sustainability approaches' the mainstream practice in the buildings and construction sector, SBC partners identified three interlinked priorities for the programme to work on, provided in Figure 2.5. These three priorities continued across the whole programme and its two selected thematic focus areas for the period 2019–2022.

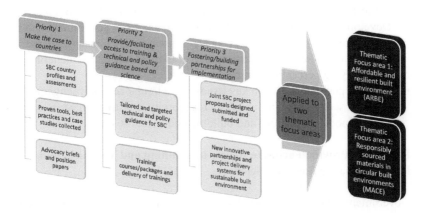

Source: SBC 2019.

Figure 2.5   *The three interlinked priorities for the SBC programme*

Priority 1: Making the Case for SBC to Countries

There is a general lack of understanding from policy makers, investors and the majority of stakeholders engaged in construction projects about the benefits of sustainable building and construction practices and the practical meaning of related concepts such as circularity, sustainably sourced materials or resilient built environment. To mobilise stakeholders in countries and secure commitment from governments and other actors, it was critical to explain what sustainable building and construction, and related concepts, mean and make the case for SBC at the national or local level by showing the benefits of SBC in their respective contexts.

Therefore, the SBC programme aimed to:

- Design and conduct targeted country assessments to analyse the current buildings and construction situation (e.g. existing policies, housing demand and related government plans, materials used and available, long-/medium-term adaptation needs/opportunities, circularity potential, costs and other related considerations) while identifying the national SBC opportunities/benefits (e.g. innovative strategies, potential innovative local material market opportunities and others)
- Collect and disseminate existing tools, best practices and case studies (e.g. decision-making tools, scenario generators, modelling tools, costing tools, financial schemes/mechanisms, territorial approaches, evaluation tools, indicators and metrics, sustainability criteria for buildings and construction projects, detailed case studies and the like) – the programme's focus was not just on documenting successes but also on learning from failures
- Develop SBC advocacy briefs for decision makers in government and the private sector (e.g. based on country assessments, proposing alternative approaches, potential impacts and the like) showing the benefits and demystifying perceptions on costs
- Develop position papers on key issues (e.g. circular economy, engaging the informal sector, materials and resource scarcity, construction waste, supply chains, model regulations and building codes, and so on).

Priority 2: Provide/Facilitate Access to Training and Technical and Policy Guidance Based on Science

Governments and other stakeholders, such as local authorities, construction workers, architects, engineers and developers, need guidance on shifting policies and actions towards the adoption of SBC practices. This guidance needs to consider local and context-related specificities. While a wealth of guidance, tools and information exists, it is often challenging for governments and other stakeholders to find and access these. The SBC programme would work with its partners to produce guidance as its signature product.

Therefore, the SBC programme aimed to:

- Use country assessments as well as the tools, best practices and case studies collected to develop guidance for countries through multi-stakeholder expert taskforces to take action or scale up actions for SBC and recommend tools and approaches recognised by experts (taking a holistic approach as much as possible) – the guidance could be tailored to specific actors; for example, policy makers, investors/banks, procurers, building entrepreneurs and others

• Promote and disseminate high-quality SBC training courses and support the development of new training packages on SBC guidance generated through the SBC programme.

Priority 3: Fostering/Building Partnerships for Implementation

To foster policy implementation in countries and further demonstrate the benefits of SBC, it was critical to support and engage stakeholders in concrete projects on the ground. By making the case, the SBC programme raised the interest and commitment of countries and local stakeholders to act for SBC and thereby identify promising project opportunities. Concrete projects give an incentive to governments not only to change or improve their policies but also to enforce them; they also foster the engagement and commitment of other stakeholders to change approaches and/or technologies and transform markets.

Therefore, the SBC programme aimed to:

• Identify needs and project opportunities in selected countries
• Identify relevant calls for proposals and facilitate the creation of consortiums to respond to tenders/requests for proposals and develop project proposals for submission to donors/funding mechanisms
• Identify opportunities to scale up successful pilots and best practices
• Provide technical guidance and support to bodies (countries, organisations) preparing tenders or project proposals.

The SBC scale-up approach is illustrated in the Theory of Change diagram (Figure 2.6).

The SBC programme continued its efforts on resource mobilisation, engagement with other OPN programmes such as the SPP programme, and communication and outreach activities.

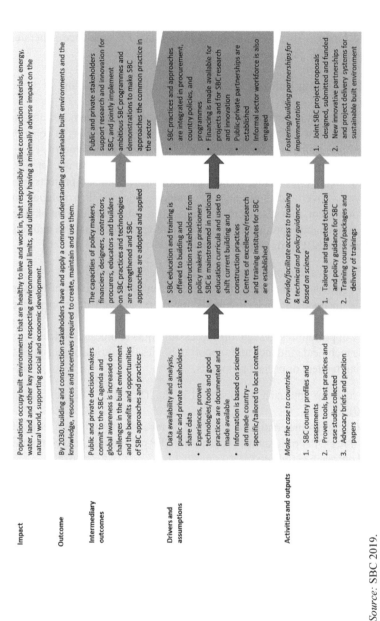

Source: SBC 2019.

Figure 2.6   SBC Theory of Change

CONCLUSION

This chapter has presented a brief history of the SBC programme. This included the structure of the programme, workplans developed to deliver on the guiding principles upon which it was set up, and the focus of the programme and its priorities as it navigated through internal and external changes. Forthcoming chapters present the value chain approach of the SBC programme and the SBC projects undertaken during the 2015–2017 period, before presenting some primary research by the programme on SDG mapping and feedback from its stakeholders.

NOTE

1. The WorldGBC moved from Co-Lead status to MAC in 2018.

REFERENCES

SBC (Sustainable Buildings and Construction programme) (2015) *Detailed Work Plan 2015–2016.*

SBC (Sustainable Buildings and Construction programme) (2019) *Sustainable Buildings and Construction Programme of the One Planet Network (One Planet SBC Programme) SBC Programme Workplan 2019–2022.* https://www.oneplanetnetwork .org/sites/default/files/sbc_work_plan.pdf

SBC (Sustainable Buildings and Construction programme) (2023) Sustainable Buildings and Construction Programme Knowledge Centre, One Planet Network. https://www.oneplanetnetwork.org/programmes/sustainable-buildings-construction/ knowledge-centre

UN-Habitat (2023) *Housing Rights Overview.* https:// unhabitat .org/ programme/ housing-rights

3. SBC value chain and actors

Pekka Huovila and Usha Iyer-Raniga

INTRODUCTION

The SBC programme worked on the circular built environment and responsibly sourced materials during the period 2019–2022, where the stages, actors and outcome in the circular construction life cycle were defined. This chapter explains the programme's view on how the construction value chain adds environmental, social and economic value in different life cycle stages involving various actors. The outcomes of the Task Group of the OPN and International Resource Panel (IRP) on the construction value chain are also presented. They include consultations on procurement, finance and planning that followed, led by UNEP.

Circular economy has been defined in many ways. The Ellen MacArthur Foundation (EMF 2020) states that circular economy is underpinned by three principles: no pollution; retaining the value of materials and products, and ensuring no waste; and allowing nature to regenerate. The United Nations Environment Assembly (UNEA) has characterised circular economy as

> one of the current sustainable economic models, in which products and materials are designed in such a way that they can be reused, remanufactured, recycled or recovered and thus maintained in the economy for as long as possible, along with the resources of which they are made, and the generation of waste, especially hazardous waste, is avoided or minimized, and greenhouse gas emissions are prevented or reduced. (UNEP/EA.4/Res.1, 2019)

It has also been described as an economic model aiming to foster sustainable economic growth, increase a company's economic value-add, boost global competitiveness and achieve the SDGs (Mackenbach et al. 2020). The World Business Council for Sustainable Development (WBCSD) states that circular economy approaches reduce environmental impacts by helping to decarbonise the construction industry and its value chain while providing a platform to improve social value impacts (WBCSD 2021). This text debates the achieved and maintained environmental, social and economic value in the construction value chain. The work on circular economy undertaken by the SBC programme has been guided by the EMF definition and the work of the WBCSD.

To understand where the most opportunities from circular processes may be nurtured, the life cycle stages need to be understood. Following the construction life cycle stages, the actors and the outcomes are described, followed by a discussion on how not just environmental but also social value may be met from the process of building and construction.

LIFE CYCLE STAGES

The International Finance Corporation (IFC 2018) defines the value chain for any construction project as being composed of specific variations within a fixed framework of distinct stages: design, production and conversion of raw materials into manufactured products, and the process of construction itself. Each of these comprise their own internal stages, processes, stakeholders and aspects that interact to bring a project to fruition. Lean practices for construction professionals describe primary value-chain activities in construction as raw materials and product manufacturing, design and engineering, construction, operation and maintenance. The other value-chain activities are demolition (we prefer the use of the term 'deconstruction' to demonstrate value retention or enhancement) of older buildings before the new construction and the renovation of a building that is already built and in use (LetsBuild 2019).

Standard EN 15643:2021 *Sustainability of construction works – Framework for assessment of buildings and civil engineering works* provides principles and requirements for the assessment of environmental, social and economic performance of buildings and civil engineering works considering their technical characteristics and functionality. It defines the construction process in Pre-Construction (A0), Product (A1–A3), Construction Process (A4–A5), Use (B1–B8) and End-of-Life (C1–C4) stages. Circularity is dealt with in benefits and loads beyond the system boundary (D), as shown in Figure 3.1 (CEN 2021).

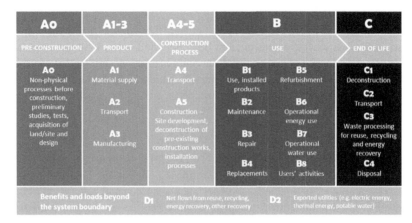

Source: Huovila and Westerholm 2022.

*Figure 3.1   Construction works assessment information in EN
              15643:2021*

Based on the international standard, general guidance, practical experience
and common knowledge, the SBC programme defined the circular life cycle
of building in six phases: Manufacturing (A1–A3), Design (A0), Construction
(A4–A5), Operation and Use (B1–B2, B6–B8), Renovation (B3–B5) and
Deconstruction/End of Life (C1–C4). These activities can be started at any
stage during the life cycle. They should not just commence with raw material
extraction and finish with producing waste, as is the case with current linear
processes. Instead, the first stage can be the operation of a building or its
deconstruction as well as design of a new or existing building, or some other
stage during the long life cycle of buildings (see Figure 3.2).

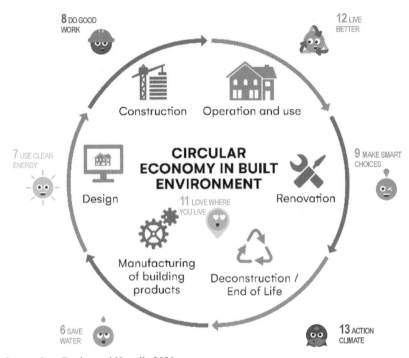

Source: Iyer-Raniga and Huovila 2020.

Figure 3.2   Construction life cycle stages

ACTORS AND OUTCOMES

When designing the actors in the value chain, the SBC programme separated renovation as an individual cycle, or number of such cycles, during the long service life of buildings, as shown in Figure 3.3. It should be noted that the various stages can also overlap, and they may include a number of iterations. As an example, design often continues during construction and renovation phases instead of remaining a separate stage to be completed before moving to the next phase. The end-users and market may also change during the process, requiring design changes during the construction process. A number of project delivery systems and their variations that are in use in defining the roles of different actors in contract modes can vary (Huovila et al. 2013, UNEP 2018). During the work of the SBC programme, the OPN and the IRP also formed a Task Group aiming to catalyse science-based policy action on sustainable consumption and production by creating the conditions to provide actionable insights on the management of natural resources in relation to the 2030 Agenda for Sustainable Development. To achieve this, a sectoral focus was taken by the Task Group and the value-chain approach applied. The benefits were demonstrated to define SCP action informed by science by applying it to three prioritised sectors: food, construction and textiles (UNEP 2021a).

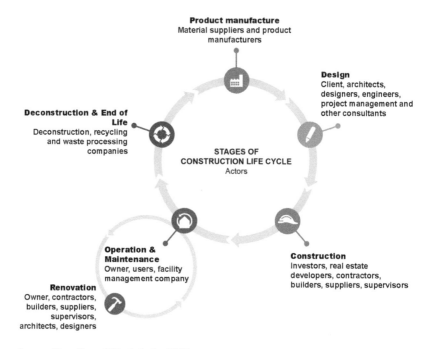

Source: Huovila and Westerholm 2022.

Figure 3.3   *Actors in different stages of a circular construction life cycle*

The value-chain analysis for construction was informed by various reports of the IRP that address aspects of construction, complemented by other sources. Based on these sources the construction value chain was visualised as shown in Figure 3.4.

Even though the textile sector in the Task Group report was presented in a circular way, construction was still illustrated in a linear way, starting from extraction of natural resources and finishing in waste disposal. The Task Group view was that this was done to highlight the current realities and challenges, and the stages of the construction value chain were shown for illustrative purposes as in Figure 3.4. The SBC Coordination Desk's view of the SBC

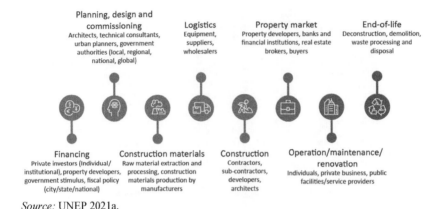

Planning, design and commissioning
Architects, technical consultants, urban planners, government authorities (local, regional, national, global)

Logistics
Equipment, suppliers, wholesalers

Property market
Property developers, banks and financial institutions, real estate brokers, buyers

End-of-life
Deconstruction, demolition, waste processing and disposal

Financing
Private investors (Individual/ institutional), property developers, government stimulus, fiscal policy (city/state/national)

Construction materials
Raw material extraction and processing, construction materials production by manufacturers

Construction
Contractors, sub-contractors, developers, architects

Operation/maintenance/ renovation
Individuals, private business, public facilities/service providers

Source: UNEP 2021a.

Figure 3.4   Stages of the construction value chain

programme from a value-chain perspective was different, and the programme chose to express this in a circular manner, due to the reasons presented below.

The use of the terminology in the Task Group document is not consistent with the buildings and construction industry. The construction process itself is a stage that is reflected in the ISO and EN standards, and contractors involved in the construction process are actors. Financing and planning stages are part of the design process (A0), influencing different stages where value is created. The property market is also influenced by financing and may be part of the A0, B and C stages in the standards as listed in Figure 3.1. Likewise, transport may occur in stages A2, B4 and C2. Logistics is not a defined stage but is interspersed across the life cycle of the building and construction process. This has been highlighted by Huovila and Westerholm (2022).

The Task Group concluded that key challenges to be addressed are what types of construction are built and used, how much is being built and how it is built (UNEP 2021a), emphasising factors that are external to the construction value chain as perceived by the SBC programme. Group members' identified intervention points were focusing on the initial stages of the construction value chain: public procurement (UNEP 2021b), financing (UNEP 2021c) and planning (UNEP 2021d) in consultative workshops (UNEP 2022).

In the SBC programme, the SDG 12 Resource Efficient Housing Project developed processes and tools to support circularity in the construction value chain (Salcedo et al. 2022). Figure 3.5 by the SBC programme compiles different views on the construction value chain, also explaining the playground outside the construction life cycle, including external actors.

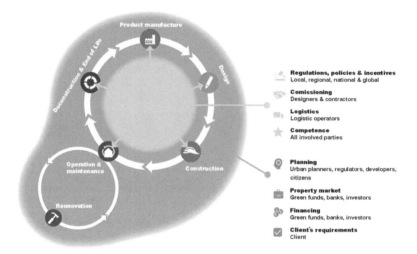

Sources: Huovila and Westerholm 2022, Salcedo et al. 2022, UNEP 2022.

Figure 3.5   External actors and playground in circular built environment

DISCUSSION

Since the construction sector has significant environmental, economic and social impacts, it is natural that it has the potential to create environmental, economic and social values. In traditional business models, value is often concentrated on the financial value for the investors and customers. In business models that are based on circular economy, value is seen more broadly as considering a wider range of value-chain partners, the environment and the society (Nußholz et al. 2020). The SBC programme illustrated the value decrease, retention and increase during the building life cycle, as shown in Figure 3.6.

Circular building requires careful planning and skilful design. Taking an element from one building and bringing it into another means identifying and working with new value chains. When new elements are made through upcycling, this creates new jobs where different sets of skills are needed compared to manufacturing new products (WBCSD 2021).

Applying circularity to construction can also produce financial value in many ways. The price of secondary materials is typically lower compared to virgin materials. Moreover, circularity helps to retain the value of construction materials and products. Circular construction is also associated with increased value to customers. Circular implementation should lower life cycle costs compared to typical linear solutions. However, sustainable construction is still perceived to have higher initial investment costs. Sustainable construction can create additional economic value by improving the competitive advantage and minimising the financial risk for investors. Buildings with low negative environmental impacts might also have positive effects on corporate image and marketing (Nußholz et al. 2020).

Circular construction can also contribute to added social value. Reusing materials is more labour-intensive and thus it creates employment. Additionally, circular construction favours repurposing and redesigning existing spaces instead of tearing them down and building anew. This allows the preservation of historical architecture and cultural heritage and identity. In many cases, the factors that contribute to environmental or economic value also contribute to added social value (Nußholz et al. 2020).

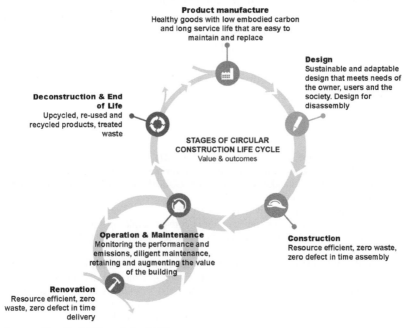

Source: Huovila & Westerholm 2022.

Figure 3.6   *Outcome and value in construction value chain*

CONCLUSION

It is critical to understand the SBC programme's approach to the circular economy. The programme clearly defined the life cycle stages in accordance with standards and accepted industry practices to shift current linear practices and adopt practices to support circularity outcomes. The approach the SBC programme took was to make it easy for the industry to seek added value in the construction value chain by enabling actors to commence their circular journey at either the manufacturing, design, construction, operation or end-of-life phases. Clearly, the production process that supported low environment impact in terms of resource use, production, operation and maintenance was preferred. Similarly, designs that are adaptable and support disassembly are valued, as are materials that provide transparency of composition. According to the SBC approach, the construction phase is warranted where resource efficiency is optimised, no waste is produced in the construction process and defects/re-work are completely avoided. The operation and maintenance phase should prolong the life of the building, no waste should be generated in operations and value of the building needs to be retained. The building's end of life should support the reuse of materials, so materials should not have any toxicity, and nor should they generate any waste.

Having outlined the understanding of the construction value chain, the next chapter details the projects that emphasise the different actors and life cycle phases of the construction process that were supported by the SBC programme, their outputs and associated achievements.

REFERENCES

CEN (Comité Européen de Normalisation) (2021) *Sustainability of Construction Works: Framework for Assessment of Buildings and Civil Engineering Works*, EN 15643:2021. https:// standards .iteh .ai/ catalog/ standards/ cen/ b98a1879 -ace4 -490b -8fa5-4f286710ae84/en-15643-2021

EMF (Ellen MacArthur Foundation) (2020) *Three Core Principles of the Circular Economy*. https://acehub.org.au/news/three-core-principles-of-the-circular-economy #: ~: text = There %20are %20three %20primary %20principlesR, egenerate%20 natural%20systems

Huovila, P., Hyvärinen, J., Mäkeläinen, T., and Huovila, A. (2013) Feasible frameworks and value models for development of value-oriented service, in M. Isomursu, M. Toivonen, M. Kokkala and P. Pussinen (eds), *Highlights in Service Research* (pp. 143–149), VTT Technical Research Centre of Finland, ISSN-L 2242-1173. https://publications.vtt.fi/pdf/researchhighlights/2013/R6.pdf#page=145

Huovila, P., and Westerholm, N. (2022) Circularity and Sustainability in the Construction Value Chain, *IOP Conference Series: Earth and Environmental Science*, 1078, 012004. https://iopscience.iop.org/article/10.1088/1755-1315/1078/ 1/012004/pdf

IFC (International Finance Corporation) (2018) *Construction Industry Value Chain: How Companies Are Using Carbon Pricing to Address Climate Risk and Find New Opportunities*. https:// www .ifc .org/ en/ insights -reports/ 2018/ construction -industry -value-chain

Iyer-Raniga, U., and Huovila, P. (2020) *Global State of Play for Circular Built Environment: A Report on the State of Play on Circularity in the Built Environment across Africa, Asia, Europe, Gulf Cooperation Council Countries, Latin America and the Caribbean, North America and Oceania – Final Report October 2020*, United Nations One Planet Network Sustainable Buildings and Construction Programme. https://www.oneplanetnetwork.org/knowledge-centre/resources/global -state-play-circular-built-environment

LetsBuild (2019) *Eliminate Waste, Increase Productivity: A Quick Guide to Lean Practices for Construction Professionals*. https://www.letsbuild.com/ebooks/quick -guide-to-lean-practices-for-construction-professionals

Mackenbach, S., Zeller, J.C., and Osebold, R. (2020) A Roadmap towards Circularity: Modular Construction as a Tool for Circular Economy in the Built Environment, *IOP Conference Series: Earth and Environmental Science*, 588, 052027. https://doi .org/10.1088/1755-1315/588/5/052027

Nußholz, J.L.K., Nygaard, F., Whalen, K., and Plepys, A. (2020) Material Reuse in Buildings: Implications of a Circular Business Model for Sustainable Value Creation, *Journal of Cleaner Production*, 245, 118546. https:// doi .org/ 10 .1016/ j .jclepro.2019.118546

Salcedo, J., Regodón, A., Duwyn, J., Morgan, G., and Decazes, C. (2022) *RE-Think Buildings: A Circular Toolkit for the Sustainability of the Housing and Construction Sector*, SDG12Hub. https:// sdg12hub .org/ sites/ default/ files/ 2022 -11/ Housing %20VAM_presentation.pdf

UNEP (United Nations Environment Programme) (2018) *Guidance Document on Procuring Sustainable Buildings and Construction*, UN and One Planet Network. https:// www .oneplanetnetwork .org/ sites/ default/ files/ guidance _document _on _procuring_sustainable_buildings_and_construction_final.pdf

UNEP (United Nations Environment Programme) (2019) *Resolution adopted by the United Nations Environment Assembly on 15 March 2019*. UNEP/EA.4/ Res.1. https://wedocs.unep.org/bitstream/handle/20.500.11822/28517/English.pdf ?sequence=3&isAllowed=y

UNEP (United Nations Environment Programme) (2021a) *Catalysing Science-Based Policy Action on Sustainable Consumption and Production: The Value-Chain Approach & Its Application to Food, Construction and Textiles*. https:// www .resourcepanel .org/ reports/ catalysing -science -based -policy -action -sustainable -consumption-and-production

UNEP (United Nations Environment Programme) (2021b) *Consultative Workshop on Sustainable Construction Value Chain Focusing on Purchasing Power: How Public Procurement Exerts Influence throughout the Construction Value Chain – Workshop Report*, UN and One Planet Network. https:// www .oneplanetnetwork .org/sites/default/files/2021-12/Construction%20Value%20Chain%20Consultations %20Workshop %20Report %20 - %20Influence %20of %20Public %20Procurement .pdf

UNEP (United Nations Environment Programme) (2021c) *Consultative Workshop on Sustainable Construction Value Chain: Follow the Investments – How Financing Shapes the Construction Value Chain: Workshop Report*, UN and One Planet Network. https://www.oneplanetnetwork.org/sites/default/files/2021-12/ Construction%20Value%20Chain%20Consultations%20Workshop%20Report%20 -%20Influence%20of%20Financing.pdf

UNEP (United Nations Environment Programme) (2021d) *Consultative Workshop on Sustainable Construction Value Chain Focusing on How Planning and Design*

Frame Action along the Construction Value Chain: Workshop Report, UN and One Planet Network. https://www.oneplanetnetwork.org/sites/default/files/2021-12/Construction%20Value%20Chain%20Consultations%20Workshop%20Report%20-%20Influence%20of%20Planning%20&%20Design.pdf

UNEP (United Nations Environment Programme) (2022) *Outcome Document: Opportunities and Gaps Identified during the Construction Value Chain Consultations*, One Planet Network. https://www.oneplanetnetwork.org/knowledge -centre/ resources/ outcome -document -opportunities -and -gaps -identified -during -construction

WBCSD (World Business Council for Sustainable Development) (2021) *The Business Case for Circular Buildings: Exploring the Economic, Environmental and Social Value*. https://www.wbcsd.org/contentwbc/download/13200/193514/1

4. SBC projects

Pekka Huovila and Usha Iyer-Raniga

INTRODUCTION

Underpinned by the guiding principles that were identified during its first years, the SBC programme was keen to demonstrate their application through research and implementation projects. The programme worked on five initiatives supported by the 10YFP Trust Fund during the period 2015–2022. The first three projects were based in Colombia, Burkina Faso, Kenya, Nepal and India and delivered during the period 2016–2019. A total of 500,472 USD in research and development on SBC was budgeted. Two more SBC Trust Fund projects followed the same theme in Egypt and in Latin America during 2018–2022 with a total budget of 167,505 USD (UNEP 2018a).

After that, a new financing mechanism was created for the OPN as the implementation platform for 10YFP programmes. SBC, acknowledged to be a high-impact sector, was awarded 500,760 USD from the One Planet Multi-Partner Trust Fund (MPTF) to complete a project in Burkina Faso and Sri Lanka involving UN organisations to implement SDG 12 on responsible consumption and production. This was undertaken at the tail end of the programme, during 2020–2022.

This chapter describes experiences from the Trust Fund process and the main results of six Trust Fund projects carried out by the SBC programme.

10YFP TRUST FUND

Each of the five 10YFP programmes was allocated around 500,000 USD from the 10YFP Trust Fund, as financial support to 10YFP initiatives in developing countries and emerging economies to run projects supporting implementation of their prioritised objectives (UNEP 2018b). General guidelines for the call for applications for funding were defined by the 10YFP Secretariat, which was managed by UNEP. Organisations that were eligible to apply were regional, national and local governmental institutions; agencies or subsidiary bodies; not-for-profits; associations; civil society organisations; community-based organisations; universities; and scientific or technical organisations with headquarters in developing countries and countries in transition. The underlying intent behind the selection of these organisations was to support specific programmes and initiatives responding to the national and regional priorities of developing countries and countries with economies in transition (10YFP Secretariat 2016).

Call for Proposals

Table 4.1   SBC Trust Fund team project proposal evaluation

Assessment criteria	Max score
Relevance of the project	*25*
Meeting the SBC programme and Call for Proposals objectives	15
Relevant to SDGs, maximising the expected impacts	10
Planning of proposed activities	*15*
Capacity building of local stakeholders	10
Geographical spread (covering different areas, countries, continents)	5
Management capacity	*20*
References from earlier projects from last 3 years	5
Technical expertise of the team to carry on the work	15
Sustainability of the project	*20*
Replicability and scaling-up potential	10
Consideration of the supply chain including SMEs (business ecosystem)	10
Budget and cost-efficiency	*20*
Is the budget realistic and cost-efficient for the proposed work?	10
Solid sources of finance	10
Total	100

Source: SBC Coordination Desk and 10YFP Secretariat 2015.

The SBC programme launched its Trust Fund Call on 3 December 2015 at the UN Climate Change Conference of Parties (COP 21) in Paris. It was prepared by the voluntary non-recused SBC Coordination Desk and SBC Multi-Stakeholder Advisory Committee (MAC) members and supported by the 10YFP Secretariat, which managed the 10YFP Trust Fund, coordinated its call processes and handled contracts with the project Leads.

   The SBC call for proposals included the following common 10YFP documents: Trust Fund Guidance on Calls for Proposals, a Trust Fund Application Form, a Trust Fund Information Form and a Trust Fund Budget Form (10YFP

Secretariat 2016). Evaluation of the submitted proposals was undertaken in two stages. The first was operated by the 10YFP Secretariat to validate their eligibility and assess how their general 10YFP criteria were met. The second phase was run by the SBC Trust Fund team that had defined SBC-specific criteria for appraisal, as shown in Table 4.1.

SBC Projects

Over a hundred (108) proposals were received from the Global South by the 4 February 2016 deadline. Of the 108, 72 proposals were judged to be eligible for funding. Finland, as the SBC Lead, conducted the preliminary assessment for 66 proposals, and UNEP and RMIT University, as the SBC Co-Leads, did the same for the remaining six. Between the 10YFP criteria and the SBC criteria, 15 applications were ranked highly. In addition, three proposals based on their highest average score from the 10YFP and SBC assessments were added to form a longlist of 18 projects. Since SBC had funds to contract only three of them directly, a closer review was conducted by the SBC assessment team, consisting of volunteer non-recused SBC network experts, to form a shortlist of the top six proposals.

The proposals were all judged to be of high quality and their review scores did not differ much. They also addressed the four main SBC work streams. Considering the global nature of the 10YFP, it was decided to look at the geographical coverage of the first projects proposed for funding, while prioritising the SBC assessment criteria. Two projects addressed the Latin America and the Caribbean (LAC) region, namely Colombia and Ecuador; two targeted Southeast Asia: India and Malaysia; one focused on Central Asia: Kyrgyzstan; and one on Africa, namely Kenya and Burkina Faso, while also stretching to Nepal in South Asia.

Based on the assessment results, the SBC Trust Fund team unanimously recommended funding the following three projects with corresponding budgets totalling 500,000 USD, as follows:

- Sustainable Construction Policy in the Aburrá Valley, Colombia: 135,000 USD
- Mainstreaming Sustainable Social Housing (MaS-SHIP) in India: 180,000 USD
- 'SHERPA' in Kenya, Burkina Faso and Nepal: 185,000 USD.

These projects focused primarily on the two first SBC themes: sustainable policies and affordable housing. In addition, they contributed to the other two SBC themes: supply chain and climate change. They were all seen to have potential for wider exploitation.

The next step was to request that these three bidding teams develop detailed project plans based on their approved proposals within the budget constraints. While these three projects started in 2016 for a duration that varied from two to three years, the SBC programme was looking for opportunities to fund runner-up projects from among the 15 shortlisted. Those starting first covered sustainable policies in Colombia, local materials for social housing in India and a design tool in Burkina Faso, Kenya and Nepal. The other top-ranking candidates addressed relevant important considerations; for example, use of bamboo in Ecuador, water resources in Namibia, recycling in Malaysia, simplified life cycle assessment (LCA) in Brazil, composite materials in the LAC region, eco-construction in Thailand and West Africa, and procurement in Argentina and Bhutan.

In 2018, the Ministry for Foreign Affairs of Finland made a pledge to the 10YFP Secretariat to start two more SBC Trust Fund projects with a smaller budget than the previous ones. Based on the additional funding, the SBC team selected the following two proposals that best met the specific SBC criteria:

- Sustainability Assessment of UpScale Smart Village and Actions for Improvement in Egypt: 83,921 USD
- Enhance the Sustainability of Concrete Production – Implementation in Latin America of the Concrete Sustainability Council Responsible Sourcing Scheme in Colombia: 83,585 USD.

In addition to these projects, the SBC programme was also supported by the Ministry of the Environment of Finland with small funds annually. These funds were used for the MAC meetings held on an annual basis, and after the COVID-19 period the funds were used to support state-of-play reports based on various geographical regions. These reports were authored by experts from the regions: Africa (Gibberd 2020), Asia (Niazi et al. 2020), Europe (Westerholm 2020), Gulf Cooperation Council Countries (Al-Alawi et al. 2020), the LAC region (Moreno 2020), North America (Keena and Dyson 2020) and Oceania (Iyer-Raniga 2020). These reports were then brought together with the *Global State of Play Report for Circular Built Environment* with key recommendations and conclusions (Iyer-Raniga and Huovila 2020). These reports were launched at the World Sustainable Built Environment Conference 2020 in Sweden. Deeper dives with case studies linked to the SDGs were undertaken in 2021 across the African, Asian and LAC regions (SBC 2023).

RESULTS

The first three SBC Trust Fund projects were run from 2016 until 2019 in three continents (Hakaste and Huovila 2019, Huovila and Iyer-Raniga 2020). Support and following-up by the SBC Coordination Desk was organised so that UNEP was looking after projects in Latin America, Finland oversaw Africa and RMIT University took care of Asia. In a second wave, two more Trust Fund projects started in 2018. The first finished in 2019 in Egypt, while the other was extended in a couple of Latin American countries until 2022 (FIGBC 2017, 10YFP Secretariat 2020).

Dissemination of the objectives and results of the Trust Fund projects was undertaken by individual projects supported by the 10YFP Secretariat, which created the One Planet Camp at the High-Level Political Forum on Sustainable Development (HLPF) in New York in 2018 for the promotion of these activities. It included a virtual reality application of the SHERPA design tool in Nepal (see Figures 4.1a and 4.1b).

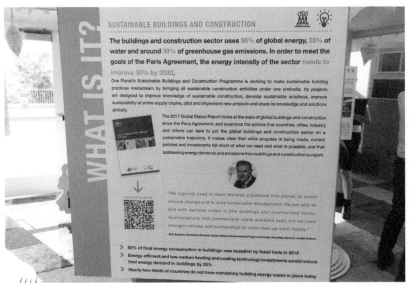

Source: © Pekka Huovila.

Figure 4.1a   *One Planet Camp at the HLPF in New York, 2018*

Source: © Pekka Huovila.

Figure 4.1b   One Planet Camp at the HLPF in New York, 2018

Sustainable Construction Policy in the Aburrá Valley, Colombia, 12 October 2016 to 31 October 2018 (10YFP Secretariat 2020)

The applicant, the state agency Área Metropolitana del Valle de Aburrá, had partners from the Colombian Chamber of Construction, Regional Autonomous Corporation of the Center of Antioquia and National Cleaner Production Center, forming a multi-disciplinary team to root sustainable policy in the region.

The objective was to foster the paradigm shift towards an eco-efficient, socially inclusive and economically viable urban development in the Aburrá Valley and its adjacent regions by implementing strategies relevant to the local needs, constraints and opportunities, while taking into account the global priorities. These conceptual, methodological and legal policy frameworks serve as a reference to issue national policy and formulate similar policies in other metropolitan regions of Latin America.

The project brought together a good partnership between the public sector, private sector and civil society using a holistic approach including a strong training component and a clear plan and commitment to maintain the activities after the end of the funded project. The approach seemed to be replicable in other urbanising metropolitan areas in the LAC region and beyond.

Stakeholder engagement included the private and public sectors. The guide-lines were promoted through, for example, Expoinmobiliaria, One-Click LCA, ExpoCamacol, Social Institute of Housing and Habitat of Medellin (see Figure 4.2), and the Municipality of Envigado. Fifty private and public entities shared their successful practices and challenges, and 165 professionals were trained in four groups. Three demonstration construction projects were a shopping centre, social housing and the administrative municipal centre of Caldas (Área Metropolitana del Valle de Aburrá and Camacol Antioquia 2019).

Source: © Pekka Huovila.

Figure 4.2   *City of Medellín is in the heart of the Metropolitan Area within Aburrá Valley*

Development of Sustainable Housing Design Tool 'SHERPA' (Kenya, Burkina Faso, Nepal), 24 August 2016 to 11 February 2019 (10YFP Secretariat 2020)

The applicant, the Kenyan government's Kenya Slum Upgrading Programme (KENSUP), had designated UN-Habitat as the implementing entity for the project. The partners were the NGOs Yaam Solidarité from Burkina Faso and Architecture Sans Frontières (ASF) Nepal, supported by French, Finnish and British research centres: the International Centre on Earthen Architecture (CRAterre), VTT Technical Research Centre of Finland (VTT) and the University of Cambridge.

The project's objective was to develop a free, easy-to-use and locally adaptable sustainable housing design tool to assist housing practitioners in developing countries in designing exemplary socially and culturally responsive, climate-resilient and economically sustainable housing projects. A core methodology with indicators and weightings was created and a first working alpha version developed. Now, extensive field testing and the development of an easy-to-use, multi-functional user interface for both online and offline applications is necessary.

A strong team was formed, developing a tool based on earlier experience that can create local guidelines and establish important benchmarks for sustainable housing. The project has huge potential for wider replicability, and has already been validated in three countries in two continents, supported by a large dissemination network provided by the core team.

The tool was developed to be used on a smartphone in English, French and Spanish to identify and analyse the strengths and weaknesses of new, current and past housing projects (UN-Habitat SHERPA n.d.). Training and testing were organised in Ecuador in 2016 before the tool was implemented in Burkina Faso, Kenya and Nepal (YAAM Solidarité 2017).

The project was introduced at Habitat III in Quito in 2016. Its results were shared at the South-South Cooperation Conference on SCP in Brasilia in 2017, the World Urban Forum (WUF 9) in Kuala Lumpur in 2018 and the Sustainable Building Knowledge Sharing Workshop in Buenos Aires in 2018.

The tool was also applied in Kalobeyei Refugee Camp in Kenya after the first pilot schemes had validated it. SHERPA was also used by CRAterre students from Algeria, Benin, Colombia, Democratic Republic of the Congo (DRC), France, Guinea, Haiti, India, Mali, Nepal, Nicaragua, Senegal and Venezuela in 2017 as part of their studies.

Mainstreaming Sustainable Social Housing (MaS-SHIP) in India, 1 October 2016 to 31 March 2019 (10YFP Secretariat 2020)

Oxford Brookes University's Low Carbon Building Group from the UK acted as the implementing entity. The project partners included Development Alternatives (DA), the Energy & Resources Institute (TERI) and UN-Habitat.

The primary focus was to enhance sustainability in the construction of social housing through adoption of suitable construction as well as operations and management practices. The tasks included facilitating sustainability through the adoption of sustainable building materials and construction technologies by social housing providers, development of a policy framework with focus on operational energy use, judicious use of material and natural resources in construction, and financial practices promoting sustainability and socio-economic considerations. Interventions in the form of suitable building materials, incorporating resource efficiency, design and different construction technologies, energy-efficient practices and financial considerations, were designed for two pilot demonstrations, awareness generation and knowledge dissemination (Herda et al. 2017).

MaS-SHIP produced a comprehensive data framework, tools, evidence-based knowledge, insights and policy recommendations for mainstreaming sustainable social housing in India. A socio-technical approach was adopted in the research, bringing together primary and secondary data collection with both quantitative and qualitative assessments, using literature review, stakeholder engagement, online and field surveys, statistical tests and thermal simulations. The construction and policy ecosystem were examined to identify barriers and opportunities in adopting sustainable building materials and related design and construction practices, so as to develop policy recommendations.

A key output is the Decision Support Toolkit comprising datasets and tools to help users in choosing sustainable building materials and developing and monitoring sustainable design and construction practices in social housing projects. The Sustainability Assessment Tool, developed in the project, has the capability to measure the relative performance of building materials and systems for social housing projects that do not exceed four stories.

Three policy briefs were developed to assist policymakers and practitioners. These involved policy measures for mainstreaming sustainable social housing, decision making using the Sustainability Index parameter, and attributes and aspects and institutional mechanisms for adopting the Sustainability Index (Gupta et al. 2018).

Sustainability Assessment of Nabta Smart Town (NST), Egypt, and Actions for Improvement, 25 January 2018 to 31 March 2019

VTT Technical Research Centre of Finland (VTT) acted as the implementing entity and Khairy Foundation for Human and Social Development (KFHSD) was involved as a project partner.

The project was aimed at sustainable improvement of the NST Master Plan in Borg El Arab, Egypt, and the different building types it contains. The work was in line with the recommendations and findings from VTT's recently concluded EcoNBC Feasibility Study. A sustainable urban model was also developed that can be locally adapted and replicated in other countries. In addition, the capacity of KFHSD staff regarding the principles and benefits of EcoCities was developed (10YFP Secretariat 2020).

The first phase focused on an EcoCity energy-efficiency feasibility assessment using a specific methodology developed by VTT in which local partner KFHSD was trained to be able to contribute to data collection, modelling and simulation, as well as system performance and optimisation. Results were shared and discussed with a selected group of students from Alexandria University and relevant stakeholders.

The main results of the energy assessment of the NST Master Plan focused on energy being imported and exported from and to the grid. On-site energy production from photovoltaics (PV) and wind was simulated to individual buildings relating to their consumption profile.

The main conclusions can be summarised as follows:

- Before responding the renewable energy system, buildings energy demand has to be reduced, so that the overall energy need itself is reduced.
- Different energy savings measures should be adopted for different building types.
- Optimising on-site consumption is necessary for reducing the cost on the energy infrastructure.
- It is possible to achieve a Net Zero Energy District (NZED) or Positive Energy District with conventional renewable energy systems (RES) (Antuña-Rozado et al. 2019).

Going beyond energy, sustainable building technologies reduce the consumption of water, which is becoming increasingly scarce in Egypt. Therefore, a follow-up study should include the sustainability assessment of the water system planned for NST. Similarly, the sustainability of the NST solid waste management plan should be assessed as solid waste represents one of Egypt's most urgent problems, both in cities and in the countryside. An array of solutions, from waste separation at the source, considering transport and logistics, and recycling, all the way to waste-to-energy or production of organic fertilisers, should be considered for implementation in NST.

The project's results were presented at the InnoFrugal Conference held in Helsinki in 2018. The SBE19 Helsinki Conference awarded the project the Best Paper Contributing to a Healthy Built Environment award from the Juho Vainio Foundation. A scientific article describing the main results has been published (Antuña-Rozado et al. 2023).

Concrete Sustainability Council (CSC) Responsible Sourcing System Implementation in Latin America, 31 January 2018 to 30 November 2022

The implementing entity was the Concrete Sustainability Council, and the project partner the Colombian Chamber of Cement and Concrete (10YFP Secretariat 2020).

The aim was to empower small and medium-sized concrete producers in Latin America to make concrete more sustainable by using training and the implementation of practical tools to help suppliers, contractors, governments and stakeholders. A number of stakeholder meetings with industry associations, green building councils and certification bodies were organised, and discussions with ministries took place to promote policies for the establishment of green public procurement in Colombia, Costa Rica and Panama.

There were no training certificates in Latin America before this project. Training and communication materials were developed in Spanish and help desk support was provided to help plants aiming to achieve CSC certification in Bolivia, Brazil, Costa Rica and Mexico. A study on informal construction in Colombia, Costa Rica and Panama was conducted (CSC and FIHP 2021).

The implementation plan of the CSC in Latin America initially focused on Costa Rica, Colombia and Panama, but has also reached Guatemala, Peru, Ecuador and Argentina. Sixteen certificates were awarded in Argentina, Colombia, Ecuador, Guatemala, Panama and Peru during the project's lifetime. More training and certification is expected to follow in the future (Lascarro 2022).

ONE PLANET MULTI-PARTNER TRUST FUND

Simultaneously with the rather small-scale Trust Fund projects, the 10YFP Secretariat developed a mechanism inviting programmes to prepare bigger flagship projects to attract various international financial institutions (IFIs), but this work was dropped after two years. The SBC programme also prepared proposals for various calls without success, such as SWITCH Africa Green,[1] SWITCH-Asia,[2] the Green Climate Fund (GCF),[3] the International Climate Initiative (IKI)[4] and Horizon 2020.[5]

A 10YFP external review undertaken in 2017 (UNEP 2018b) made similar observations as the six programmes, in that the Trust Fund needed additional resources to support successful implementation of its programmes. The importance of the formation of a solid financial basis for the programmes by the 10YFP was identified as a need to achieve expected impacts.

A fundamental change is needed in the way we consume and produce to protect the planet. SCP is one of the most cost-efficient and effective ways to achieve economic development, reduce impacts on the environment and advance human well-being (OPN 2018). To address this need, a new funding mechanism supporting SDG 12 was created supporting these objectives.

New Funding Mechanism

New partner organisations were brought together to support UNEP's efforts in resource mobilisation. As mentioned above, the MPTF was established in 2018 and launched at the HLPF. The MPTF is a partnership between six UN agencies engaged in the One Planet Network: UNEP, the Food and Agriculture Organization (FAO), the UN World Tourism Organization (UNWTO), UN-Habitat, the UN Office for Project Services (UNOPS) and the UN Development Programme (UNDP), administered by the UN Multi-Partner Trust Fund Office (FAO et al. 2021).

This Fund is a pooled funding mechanism for cohesive action to support countries in assessing, designing, prioritising and implementing policies and practices for SCP, bringing together the expertise of the OPN members and the comparative advantages of the participating UN agencies.

With the limited financial resources available, the MPTF pilot programmes aim to demonstrate the potential of the Fund to strengthen an inter-agency delivery model to countries that builds on the global pool of knowledge, resources and partners such as the OPN and brings it to the national level in a way that serves the needs of the countries and responds to their priorities (FAO et al. 2020, OPN 2018). See Figure 4.3.

The Fund was designed to enhance coordination for a system-wide delivery for the implementation of SDG 12, including at national level in the context of the reform of the UN Development System. It supported the implementation of the SBC inter-agency project in two pilot countries with the participation of the three UN agencies.

SDG 12 Resource Efficient Housing in Burkina Faso and Sri Lanka: Multi-Agency Support to UN Country Teams in Mainstreaming Resource Efficiency in the Housing, Buildings, and Construction Sector, 5 October 2020 to 31 December 2022

The convening agency was UN-Habitat, with other UN organisations (PUNOs), UNEP and UNOPS participating. The partners were national authorities from Burkina Faso and Sri Lanka supported by UN Country Teams (UNCTs). Sri Lanka and Burkina Faso were selected for implementation based on the methodology structured around three pillars: government interest and commitment, UN Country Office interest and commitment, and data and information availability (FAO et al. 2021).

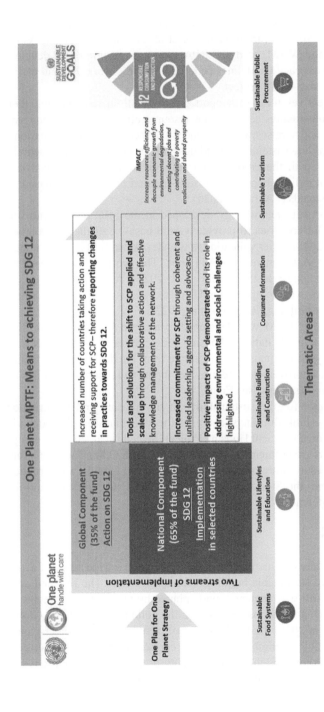

Source: Reproduced with permission from OPN 2018.

Figure 4.3   One Planet Multi-Partner Trust Fund

Results from earlier SBC Trust Fund projects including SHERPA and MaS-SHIP were explored further (Lalande et al. 2020). The Housing Value Assessment Method (VAM) used SHERPA as a starting point for the development and SBC Advocacy Guidelines benefited from the work that had earlier been done in the MaS-SHIP project.

The main objectives were to support governments in transforming their buildings and construction sector to be resource-efficient, low-carbon and climate-resilient, and to raise awareness and strengthen the capacity of national stakeholders regarding sustainable buildings and construction-related benefits and opportunities with the emphasis on material efficiency. The approach involved coherent policy support and results-focused programming based on science and local needs (FAO et al. 2020).

Country assessment methodology developed in the SBC programme was tailored to the pilot countries to provide a solid baseline of the construction sector (UN et al. 2022). After validation, it was developed further as a Country and Construction Assessment (C-CAT) tool. The work was followed by national SBC roadmaps (Hettiarachchi et al. 2022). A Mainstreaming and Advocacy Package was compiled for UNCTs, covering policy, investment value and capacity building (Regodón et al. 2022).

Finally, a housing VAM was developed to analyse the value of shifting towards more sustainable national housing programmes or large projects. It supports the analysis of a housing programme from the design stages, identifying the gaps for improvement, and provides inputs and recommendations on how to strengthen these gaps (see Figure 4.4).

A circular toolkit for the sustainability of the housing and construction sector, called RE-Think Buildings, compiles the tools that were developed in the SDG 12 Resource Efficient Housing Project (Salcedo et al. 2022). It equips UNCTs to support governments in transforming their buildings and construction sector. The results were presented at the Africities Circular Construction and Housing event, the Stockholm+50 meeting, WUF 11 Katowice and the SBEfin2022 Conference.

DISCUSSION

The SBC programme commenced in 2015, working on a wide range of activities in the field of sustainable buildings and construction. The first and only SBC Trust Fund call was launched, underpinned by those general themes. It showed a great demand for sustainability in the building sector in the Global South. However, in 2016, the funds were sufficient to contract directly only the best three out of 72 eligible proposals, representing less than 5 per cent of the potential projects. Eighteen high-quality proposals were shortlisted to be contracted next, once additional funding was available. Only two of them

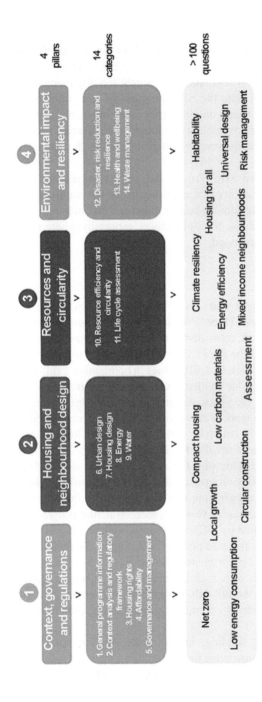

Source: Salcedo et al. 2022.

Figure 4.4   Four pillars of a housing VAM analysis

finally got started two years later in 2018 with a smaller budget, with support from Finland.

The independent 10YFP expert review observed in 2017 that the 10YFP was dramatically resource-constrained, stating that a comprehensive review of the Trust Fund was necessary. Action was taken and a new funding mechanism was launched in 2018. At that time, the SBC programme had strategically decided to work on a narrower scope on circularity and responsibly sourced materials. The next SBC call for proposals was planned to be directed specifically to these topics. Sadly, no funds were ever available to launch such a new call.

On a positive note, one more SBC Trust Fund project started in 2020, financed by the MPTF. However, its scope was defined by the UN Multi-Partner Trust Fund Office on SDG 12, sustainable consumption and production, which did not fully align with the strategic SBC objectives, especially where other funding sources needed to be canvassed.

Resource Mobilisation

During its lifetime of almost eight years, there was no basic funding to run the SBC programme's activities. Three Coordination Desk members were required to allocate one full-time equivalent for the coordination activities. The Government of Finland and the RMIT University managed to fulfil their promise of support. UNEP, however, had difficulties in nominating a full-time person to the Coordination Desk. One challenge was that the nominated person from UNEP was changing over the years. Around 20 MAC experts provided their in-kind contribution to the work with varying resources. Over 100 SBC network members from almost 50 countries gathered to provide local knowledge in the countries prepared to be involved in implementation.

Over time it became clear that despite the ambitious plans of the SBC programme, only funded activities progressed and delivered results, at least in the Global South where the programme was focused. Since the resource mobilisation by the 10YFP Secretariat was not successful, the programme prepared tenders for calls launched by different donors. Such call topics were often very specific and did not emphasise primary SBC targets. Preparation of tenders was time-consuming, and unfortunately the success rate remained very low. Enthusiasm and willingness were high but major donor funding requires specialist expertise. In addition, the full SBC programme was not involved in fundraising; only a small number of committed programme members made the effort.

Despite the issues in raising funding, luckily some small internal seed funding was obtained to build knowledge that could be applied in Trust Fund projects. As an example, country assessment methodology that was developed in SBC, supported by UNEP seed funding, was piloted in the MPTF's SDG 12 Resource Efficient Housing Project in two countries and developed further as a C-CAT tool, validating it as ready to be used in all new countries where implementation was prepared. Also, circular built environment (CBE) reports and their related 2030 Agenda indicators were made available in the same project, based on a combination of in-kind work and a small seed fund from the Ministry of the Environment of Finland.

Impacts

All SBC Trust Fund projects were aimed at mitigating emissions and improving resilience in the built environment while simultaneously supporting creation of local jobs and business opportunities. The selection of five SBC Trust Fund projects emphasised their replicability and scaling-up potential considering the local business ecosystem so that they would not remain as standalone demonstrations. That was achieved only to some extent in the short term. Luckily, continuation of SBC project activities has been assured after the initial governance structure of SBC that was formed in 2015 came to an end in 2022. Ongoing funds were secured by SBC programme members until 2025 with an objective to continue to develop extended activities towards 2030 in some form.

Active dissemination activities at major international events, such as COP climate meetings, WUFs, Habitat III or sustainable built environment (SBE) conferences, have supported awareness raising. In addition, a number of videos have been produced and webinars organised. SBC Trust Fund project findings have also been documented in research reports and scientific publications.

Selected examples of the achieved impacts are as follows. All Trust Fund projects had a strong training component building the capacity of different stakeholders in the value chain. The SHERPA tool was developed further in the MPTF project as a housing VAM and can be applied to upcoming projects. The MaS-SHIP Decision Support Toolkit provided a basis for mainstreaming and advocacy guidelines for the same project. Sustainable policies in Aburrá Valley and the energy simulation in NST planning can be applied in other neighbourhoods, cities or regions. CSC certification started in three countries, spreading to six, with high expectations of broadening to still more countries and more certificates after the project finished.

Changes in the building sector do not happen fast, and quantified impacts are hard to measure in buildings that commonly have a long service life. Results achieved in the five 10YFP Trust Fund projects in India, Nepal, Egypt, Kenya, Burkina Faso and Colombia were brought into the One Planet MPTF project for local adaptation and further development in Sri Lanka and Burkina Faso.

Bringing the knowledge to the Transforming the Built Environment through Sustainable Materials project that started in 2022, funded by BMZ Germany, with additional external funding, provides the opportunity to implement the knowledge and tools to be implemented in India, Bangladesh, Ghana and Senegal until 2025. There are plans to ensure a continuum of project activities based on this work until 2030 and even beyond.

CONCLUSION

In the period from 2015–2022, the SBC programme supported a total of six Trust Fund projects in the Global South. Three of them started in 2016 and were rolled out in Colombia, Burkina Faso, Kenya, Nepal and India based on the programme's call for proposals in 2015. With a budget of 500,472 USD from the 10YFP Trust Fund, three projects were planned and delivered. Two more started in 2018 in Egypt and in Latin America, supported by 167,505 USD after a donation received from Finland. In 2020, the OPN MPTF awarded 500,760 USD to the SBC project focusing on SDG 12 in Burkina Faso and Sri Lanka.

Results of the 10YFP Trust Fund projects have been implemented in the partner countries where their scaling up has remained somewhat of a challenge. They have, however, been in use in the OPN MPTF project and developed further to be applied in follow-up projects, focused on housing in particular. One of them, the BMZ-funded project, started in 2022, to be run until 2025 with external funding realised in India, Bangladesh, Ghana and Senegal, supported by the German government. Such types of empirical projects provide an opportunity for the work developed by the SBC programme to be continued beyond the duration of the programme.

NOTES

1.   https://www.unep.org/switchafricagreen/
2.   https://www.switch-asia.eu/
3.   https://www.greenclimate.fund/
4.   https://www.international-climate-initiative.com/en/
5.   https://research-and-innovation.ec.europa.eu/funding/funding-opportunities/ funding-programmes-and-open-calls/horizon-2020_en

REFERENCES

Al-Alawi, B., Cavallari, G., Cindric, H., Gargallo, M., Hegazy, N., Saqib, S., Shaka, H., and Tibi, G. (2020) *State of Play for Circular Built Environment in Gulf Cooperation Council: Sustainable Buildings and Construction Programme Regional Report for Africa – Final Report May 2020*, ARUP and United Nations One Planet Network

Sustainable Buildings and Construction programme, ISSN: 978-952-361-395-9. https://www.oneplanetnetwork.org/sites/default/files/from-crm/gcc_final_210214.pdf

Antuña-Rozado, C., García-Navarro, J., and Reda, F. (2019) Smart and Sustainable Urban Development in Egypt: The Case of Nabta Smart Town, *IOP Conference Series: Earth and Environmental Science*, 297, 012002. https://iopscience.iop.org/article/10.1088/1755-1315/297/1/012002/pdf

Antuña-Rozado, C., García-Navarro, J., and Reda, F. (2023) Early Energy Assessment at the Neighborhood Level to Promote Greater Energy Efficiency: The Case of Nabta Smart Town in Egypt, *Habitat International*, Vol. 136, 102782. https://doi.org/10.1016/j.habitatint.2023.102782

Área Metropolitana del Valle de Aburrá and Camacol Antioquia (Cámara Colombiana de la Construcción) (2019) *Implementation of the Policy for Sustainable Construction in Valle de Aburra* (video). https://vimeo.com/user10417749/review/387438295/6f4ed75055

CSC and FIHP (Concrete Sustainability Council and Federación Iberoamericana del Hormigón Premezclado) (2021) *Update Report: Concrete Sustainability Council – Responsible Sourcing System for Concrete*, One Planet Network. https://www.oneplanetnetwork.org/sites/default/files/from-crm/One%2520Planet%2520Update%25202021%2520CSC_0.pdf

FAO, UNDP, UNEP, UN-Habitat, UNOPS and UNWTO (Food and Agriculture Organization, UN Development Programme, UN Human Settlements Programme, UN Office for Project Services and UN World Tourism Organization) (2020) *One Planet Multi-Partner Trust Fund for SDG 12 Report 2020*, One Planet Network. https://www.oneplanetnetwork.org/sites/default/files/from-crm/mptf_for_sdg_12_annual_report_2020.pdf

FAO, UNDP, UNEP, UN-Habitat, UNOPS and UNWTO (Food and Agriculture Organization, UN Development Programme, UN Human Settlements Programme, UN Office for Project Services and UN World Tourism Organization) (2021) *One Planet Multi-Partner Trust Fund for SDG 12 Report 2021*, UNDP. https://mptf.undp.org/sites/default/files/documents/One%20Planet%20MPTF%20for%20SDG%2012_%20Report_2021_FINAL.pdf

FIGBC (Green Building Council Finland) (2017) *Trust Fund Final Video*. https://vimeo.com/387437474/b337c9e2a4

Gibberd, J. (2020) *State of Play for Circular Built Environment in Africa: Sustainable Buildings and Construction Programme Regional Report for Africa – Final Report May 2020*, Gauge and UN One Planet Network Sustainable Buildings and Construction programme, ISSN: 978-952-361-260-0. https://www.oneplanetnetwork.org/sites/default/files/from-crm/africa_final_201030.pdf

Gupta, R., Seth, S., Niazi, Z., Villanueva, J., Tuteja, S., Behal, M., Caleb, P., and Banerjee, A. (2018) *Mainstreaming Sustainable Social Housing in India: Findings and Insights from the MaS-SHIP project*, Mainstreaming Sustainable Housing, ISBN: 978-1-9165043-0-1. https://www.mainstreamingsustainablehousing.org/_files/ugd/caf2de_c8cff108d8644b47acebf66d27fea0b0.pdf

Hakaste, H., and Huovila, P. (2019) Mainstreaming Affordable Sustainable Housing Practices in Developing Countries, SBE19 Helsinki Conference, Helsinki, 22–24 May 2019.

Herda, G., Rani, S., Caleb, P., Gupta, R., Behal, M., Gregg, M., and Hazra, S. (2017) *Sustainable Social Housing in India: Definition, Challenges and Opportunities – Technical Report*, Oxford Brookes University, Development Alternatives, The

Energy and Resources Institute and UN-Habitat, ISBN: 978-0-9929299-8. https://www .m ainstreami ngsustaina blehousing .org/ _files/ ugd/ caf2de _82 9d1d976a8c 42f989c8b03b7de9d0e0.pdf

Hettiarachchi, M., Munasinghe, J.M., Samarakoon, D., Usuf, A., and Aleem, M. (2022) *Sri Lanka Sustainable Housing and Construction Roadmap: 2020–2050*, Ministry of Urban Development and Housing. https://globalabc.org/sites/default/files/2022-11/ Sri %20Lanka %20sustainable %20housing %20and %20construction %20roadmap .pdf

Huovila, P., and Iyer-Raniga, U. (2020) Mainstreaming Sustainable Housing for All, *IOP Conference Series: Earth and Environmental Science*, 588, 042006. https:// iopscience.iop.org/article/10.1088/1755-1315/588/4/042006/pdf

Iyer-Raniga, U. (2020) *State of Play for Circular Built Environment in Oceania*: *Sustainable Buildings and Construction Programme Regional Report for Oceania – Final Report April 2020*, RMIT University and UN One Planet Network Sustainable Buildings and Construction programme, ISSN: 978-952-361-399-7. https:// www .oneplanetnetwork.org/sites/default/files/from-crm/oceania_final_201101.pdf

Iyer-Raniga, U., and Huovila, P. (2020) *Global State of Play for Circular Built Environment: A Report on the State of Play on Circularity in the Built Environment across Africa, Asia, Europe, Gulf Cooperation Council Countries, Latin America and the Caribbean, North America and Oceania – Final Report October 2020*, United Nations One Planet Network Sustainable Buildings and Construction Programme. https://www.oneplanetnetwork.org/knowledge-centre/resources/global -state-play-circular-built-environment

Keena, N., and Dyson, A. (2020) *State of Play for Circular Built Environment in the USA: Sustainable Buildings and Construction Programme Regional Report for the USA – Final Report May 2020*, Yale CEA and UN One Planet Network Sustainable Buildings and Construction programme, ISSN: 978-952-361-395-9. https:// www .oneplanetnetwork.org/sites/default/files/north_america_final_201101.pdf

Lalande, C., Salcedo, J., Regodón, A., Duwyn, J., Morgan, G., and Bajpai, A. (2020) *Inception Report: SDG12 Resource Efficient Housing*, One Planet Multi-Partner Trust Fund for SDG 12. https://www.oneplanetnetwork.org/sites/default/files/from -crm/200406_Inception%2520report_Complete_compressed.pdf

Lascarro, M. (2022) Responsible Sourcing for Concrete as a Construction Material in Latin América, SBE22 Helsinki Conference, Helsinki, 23–25 November 2022.

Moreno, P. (2020) *State of Play for Circular Built Environment in Latin America and the Caribbean (LAC): Sustainable Buildings and Construction Programme Regional Report for LAC – Final Report May 2020*, Ecosur Network and UN One Planet Network Sustainable Buildings and Construction programme, ISSN: 978-952-361-268-6. https:// www .oneplanetnetwork .org/ sites/ default/ files/ from -crm/lac_final_201101.pdf

Niazi, Z., Singh, A., and Sen, I. (2020) *State of Play for Circular Built Environment in Asia: Sustainable Buildings and Construction Programme Regional Report for Asia – Final Report April 2020*, TARA/Development Alternatives and UN One Planet Network Sustainable Buildings and Construction programme, ISSN: 978-952-361-264-8. https:// www .oneplanetnetwork .org/ sites/ default/ files/ from -crm/asia_final_201101.pdf

OPN (One Planet Network) (2018) *One Planet Multi-Partner Trust Fund for SDG 12 Terms of Reference 2018–2022*. https:// www .oneplanetnetwork .org/ sites/ default/ files/2021-10/one_planet_mptf_tor_final_0.pdf

Regodón, A., Salcedo, J., Morgan, G., Gupta, M., Thanduparakkal, S., and Celenteno, G. (2022) A Comparative Analysis of Sustainable Construction Tools and Their Applicability to the Housing Sector in Two Countries; Burkina Faso and Sri Lanka, *IOP Conference Series: Earth and Environmental Science*, 1122, 012051. https://iopscience.iop.org/article/10.1088/1755-1315/1122/1/012051

Salcedo, J., Regodón, A., Duwyn, J., Morgan, G., and Decazes, C. (2022) *RE-Think Buildings: A Circular Toolkit for the Sustainability of the Housing and Construction Sector*, UN and One Planet Network. https://sdg12hub.org/sites/default/files/2022-11/Housing%20VAM_presentation.pdf

SBC (Sustainable Buildings and Construction programme) (2023) *UN OPN SBC Knowledge Centre*, UN One Planet Network Sustainable Buildings and Construction programme. https://www.oneplanetnetwork.org/programmes/sustainable-buildings-construction/knowledge-centre

SBC Coordination Desk (Sustainable Buildings and Construction programme) and 10YFP Secretariat (10-Year Framework Programme) (2015) *Call for Proposals for the 10YFP Sustainable Buildings and Construction Programme.*

10YFP Secretariat (10-Year Framework Programme) (2016) *10YFP Trust Fund Internal Process Guidelines 2016*, One Planet Network. https://www.oneplanetnetwork.org/sites/default/files/10yfp_trust_fund_process_internal_guidelines_final.pdf

10YFP Secretariat (10-Year Framework Programme) (2020) *10YFP Trust Fund Report 2020*, One Planet Network. https://www.oneplanetnetwork.org/sites/default/files/trust_fund_report_2020_final.pdf

UN, WWF (World Wide Fund for Nature) and Ministry of Urban Development and Housing Sri Lanka (2022) *Sustainable Building Construction (SBC) Country Assessment Report: Sri Lanka*, SDG 12 Resource Efficient Housing Project. https://envirodm.org/wp-content/uploads/2023/04/Sri-Lanka-Sustainable-Housing-and-Construction-Roadmap_2020_2050-_high-res.pdf

UNEP (United Nations Environment Programme) (2018a) *10YFP Trust Fund Report 2018*, UNEP and One Planet Network. https://www.oneplanetnetwork.org/sites/default/files/from-crm/trust_fund_report_2018_final_updated.pdf

UNEP (United Nations Environment Programme) (2018b) *10-Year Framework of Programmes on Sustainable Consumption and Production Patterns (10YFP) Independent External Review 2017, Final Report*, UNEP and One Planet Network. https://www.oneplanetnetwork.org/sites/default/files/ier_final.pdf

UN-Habitat SHERPA (United Nations Human Settlements Programme) (n.d.) *Your Guide to Sustainable Housing.* https://unhabitat.org/initiative/sherpa-your-guide-to-sustainable-housing

Westerholm, N. (2020) *State of Play for Circular Built Environment in Europe: Sustainable Buildings and Construction Programme Regional Report for Europe – Final Report June 2020*, UN One Planet Network Sustainable Buildings and Construction programme, ISSN: 978-952-361-391-1. https://www.oneplanetnetwork.org/sites/default/files/from-crm/europe_final_201030.pdf

YAAM Solidarité (2017) *SHERPA YAAM Solidarité* (video). https://vimeo.com/user10417749/review/387438259/ba016542bb

5. SDG mapping

Usha Iyer-Raniga

INTRODUCTION

This chapter presents the primary research undertaken by the SBC programme on mapping against the SDGs. The SBC was the first international programme that focused on circular built environments, as explained in the previous chapters. The research undertaken on circular built environments was also a world first.

PROGRESS ON THE SDGS

The SDGs came into effect on 1 January 2016, with a timeline for operation of 2016–2030 (UN 2023b). They consist of 17 goals, 169 targets and 248 indicators. The SDGs have been reviewed in depth at the HLPF since 2016. The progress of each SDG towards 2030 has also been reported annually since 2016. Over the last three years, the review of the SDGs showed that some improvements have been made, but many have not. Reducing inequality, lowering carbon emissions and tackling hunger stalled in 2021 (UN 2021), and the global extreme poverty rate rose for the first time in over 20 years. In the 2022 report (UN 2022), a similar trend appears as people and countries struggled in the aftermath of the COVID-19 crises, in addition to dealing with climate change, which impacts food and nutrition, health, education, the environment, peace and security, affecting all the related SDGs. The report details the reversal of years of progress in eradicating poverty and hunger, improving health and education, providing basic services, and much more. It also points out areas that need urgent action to rescue the SDGs and deliver meaningful progress for people and the planet by 2030. The year 2023 marked the halfway point of the SDGs, and this was encapsulated in the *Global Sustainable Development Report* (GSDR) (UNEP 2023).

The need for monitoring and tracking SDG progress is critical. Voluntary national reviews are helpful in tracking such progress. All member states of the UN (193 states) need to commit to peaceful cooperation in the service of the SDGs. For instance, rather than financing defence, more funds should be committed to climate change and the SDGs. Targets should be set to other deadlines, such as 2050, to accelerate progress on the SDGs beyond the current timeframe of 2030.

The progress at the midway point of the SDGs has shown that all the SDGs are off track to meet the timeframe of the 2030s. According to the GSDR (UNEP 2023), there is even greater urgency to build momentum and speed up the progress with the SDGs and ensure that time and resources are used effectively and efficiently to take a systematic and systemic approach to drive and accelerate transformations to realise the SDG Agenda. While there has been some progress overall, there has been very little work undertaken to date to ensure ongoing progress. In 2019, the then GSDR suggested that governance, economy and finance, science and technology, and individual and collective action be used to drive the progress on the SDGs (UNEP 2023). Limited progress has been made on SDGs 12, 13, 14 and 15, and this lack of progress has been largely attributed to the COVID-19 pandemic (Sachs et al. 2023). Compared to 2018/2019, progress is faster in increased research and development funding (9.5.1), safe and affordable housing (11.1.1) and implementation of official development assistance (ODA) commitments (17.2.1) (UNEP 2023). On the other hand, reducing global GHG emissions (13.2.2) and preventing the extinction of species (15.5.1) have continued to regress (UNEP 2023).

The Sustainable Development Solutions Network (SDSN) estimates that, on average, only 18 per cent of the SDG targets are on track to be achieved by 2030. These on-track targets are related to health and infrastructure services such as access to bank accounts and mobile use. Progress has been made globally on SDGs 6, 7 and 9, but progress on biodiversity, urban pollution, housing, strong institutions and peaceful societies is struggling. This, of course, varies from country to country. The SDG Index is designed to provide equal weight to each Goal. There are 97 global indicators, and the data comprises statistics collated by UN agencies and others from elsewhere. These indicators have been peer-reviewed (Sachs et al. 2023). The country that has scored the most is Finland (86.8), followed by the US and Australia with an equal score of 75.9. China is ranked at 63 (score of 72) and India at 110 (score of 63.4), and South Sudan, with the lowest score of 38.7, is ranked 166. Unsurprisingly, the Nordic countries and the EU member states are committed to the SDGs, and African countries are at the tail end of the SDGs. Many SIDS (Small Island Developing States) were not included due to unavailability of data (Sachs et al. 2023).

To ensure that the work stays on track to meet the aims of the SDGs, the current investment patterns and overall volumes of investment need to be considered. A new global financial architecture model is critical to ensure that work on the SDGs is stimulated. In addition to financial support, data literacy is needed. Sachs et al. (2023) also highlight the importance of sustainable development pathways, including the need to move to zero-carbon/net zero emissions, and the need to set up sustainable cities in a growing urbanised world that is also facing the problems associated with climate change.

Six long-term inter-related plans for investment to realise the SDGs have been suggested by the SDSN. These involve: universal quality education and supporting an economy that is underpinned by innovation, universal health, zero-carbon energy systems, sustainable ecosystems, agriculture and climate resilience, sustainable cities and transformation of digital access and services (Sachs et al. 2023).

THE BUILT ENVIRONMENT AND THE SDGS

The built environment contributes to our overall carbon impacts. Yet, there has not been a great deal of empirical research undertaken to link the built environment with the SDGs. The property and construction sector has an essential role to play in facilitating and achieving the SDGs. A shift within the industry is needed to ensure widespread achievement. Collaboration is critical and, supported by policy and strategies, health, education, economic growth, and mitigative and adaptive measures, may be successfully applied globally (BRE n.d.). There needs to be alignment across countries and organisations and at a project level.

Academic research has increased in recent times; however, there are some inherent contradictions. For instance, with the use of advanced building materials there is a contradiction with SDG 1 on no poverty, as costs to use such types of building materials are prohibitive, which will make it difficult to involve local labour. If building materials have some form of toxicity, it is also difficult to reuse them, which in turn has an impact on the manufacturing and use processes. These in turn impact SDGs 12 and 3. Indana and Pahlevi (2023) report that there are as many studies on SDGs in wealthy nations as there are in poor nations. The economic and social effects of the SDGs are the main areas of focus amongst the SDGs. This is followed by exploratory and case studies on the application of the SDGs. The theoretical and policy implications of the SDGs need to be considered carefully so they are not contradictory.

Academic Literature

In a paper by Trane et al. (2023), it was found that academic literature is largely focused on the environmental underpinnings of sustainability – SDGs 6, 7, 12, 13 and 15 – whereas more attention needs to be devoted to SDGs 4, 5 and 10. The most common fields of research in the EU were governance, CE, ecosystem services, urban localisation and decision making. In other research, by Alfirević et al. (2023), it was found that the US and the UK had the highest amount of scientific research when considering the SDGs. The greatest interest in research was in the period between 2017 and 2022, and since then the numbers have declined. The authors state that SDG research probably peaked in 2018, but the importance of international collaboration and trans-disciplinarity are the main characteristics of the SDG-focused research. The role of universities in supporting SDG 4 remains critical. Along similar lines, Chahdi et al. (2023) state that universities' activities and decisions should engage with sustainability, and SDG practices should be integrated in their own operations.

Tichenor et al. (2022) discuss the role of epistemic infrastructure across three levels: data and indicators, termed as materialities; interlinkages, such as networks and communities; and paradigms as new ways of doing work. They use the SDGs as an example of producing governing knowledge. They argue that infrastructures are politically produced and can have significant impact on social effects, which may or may not be anticipated.

Norouzi et al. (2021) undertook a bibliometric analysis of CE in the buildings and construction sector and showed that the work is focused on three main clusters. These comprise energy and energy efficiency in buildings; recycling, waste management and alternative construction materials; and sustainable development. Research amongst scholars has been mainly concerned with the development and use of alternative construction materials, circular business models, and the use of IT and smart tech through smart cities and Industry 4.0. Wieser et al. (2019) undertook a systematic literature review of systemic approaches with the SDGs and used visualisation to understand trade-offs so that decision making from a holistic perspective is facilitated.

Both emerging economies and highly developed economies have struggled with the SDG paradigm in the buildings and construction sector. In a study based in China, Liu et al. (2023) show that when machine learning was applied across 297 Chinese cities to identify patterns of CO_2 emissions, economic growth and employment structure, it was found that the industrial transitions are not always optimal. Manufacturing and construction industries with sustainable technical inputs can support decoupling and also stabilise social employment. City-level interactions between SDGs 8 and 13 in particular can spur sustainability improvements across the country. Continuing on the theme of cities and infrastructure, green infrastructure in the Indian context undermines global biodiversity and ecological services. In greenfield urban development projects, the planning and design of low-carbon buildings, coupled with green infrastructure, assists in supporting better health, while addressing other considerations such as water, food and agriculture, climate change, biodiversity and other such critical considerations regarding the ecosystem (Sharma and Kharbanda 2023).

In another example of an emerging economy, in India green infrastructure is beset with a lack of standardisation, limited knowledge, insufficient funds and resources, a lack of interconnectedness between SDGs and biodiversity, a lack of community engagement and participation, a lack of education, uncertainties associated with climate change and complexity in the various infrastructure systems (Sharma and Kharbanda 2023). In another study focused on India, Kandpal and Okitasari (2023) focused on the localisation of SDG 11 in India by undertaking a systematic evaluation of the national-level changes. These were then used to examine sub-national responses. Their findings demonstrate that the governance transformation for SDG 11 localisation in India is positive. However, there is still a need to generate deep transformational reforms given its developing stages. Practical approaches to ensure these changes proceed are still lacking.

In research undertaken by Ido et al. (2023) focusing on Brazil and the construction sector, it was found that promoting and encouraging sustainability-focused events, education on sustainability, water treatment and community project developments all supported sustainability outcomes. The associated Goals that were prioritised were SDGs 1, 4, 7, 9, 11 and 12. Hussain et al. (2023) show that general considerations of cost, quality, resource conservation and sustainability are key issues across all construction industries, not just in the case of Pakistan, which is struggling to adopt sustainability and quality-oriented practices. In this context, their research shows that the most significant driving enablers are government support and incentives, availability of financial resources, top management support to enable sustainability outcomes, organisational capability and quality material levels.

As stated, the role of infrastructure also needs to be considered alongside buildings. And, in this context, the role of building materials in the sector is critical. The built environment has adverse environmental impacts, and it also impacts on future resources. For instance, if Hanoi in Vietnam is to continue its growth in development, then the material demand to meet the needs of its infrastructure plan is far higher than the historical consumption trends. This has an impact on the resources needed to meet the plan, but also from the perspective of the needs of the future (Bide et al. 2023). In another instance, regarding Singapore, An (2023) argues for a robust and rigorous approach to sustainability reporting that incorporates value and impact for long-term resilience and business growth. She demonstrates an example of the development of sustainability reporting trends, and the benefits and challenges of using the Global Reporting Initiative (GRI) as a core reporting framework in an evolving environmental, social and governance (ESG) reporting landscape from the perspective of a real estate company, City Developments Limited (CDL). Using this as a case study, An (2023) shows that for effective sustainability performance, reporting is critical. Both value and impact may be gained through the process.

In a study presented by Wiik et al. (2023) for the Norwegian context, it was found that many of the SDGs did not correlate with the classifications of the building sustainability schemes. Most schemes responded to SDG 6. All schemes have an impact on SDGs 3, 7, 9, 11, 12, 13 and 15. On the other hand, research by Wen et al. (2020) shows that green buildings can be used to evaluate against the SDGs. Results showed that SDGs 3, 7, 11 and 12 are the Goals that most benefit from green buildings, with SDG 12 benefiting the most. SDG 7 has the most promise due to the focus of green buildings on energy.

A Dutch study (Gade and Opoku 2020) focusing on SDGs and building owners showed that the Goals that were highly prioritised were SDGs 7, 8, 11, 12 and 13, with 94 per cent of the building owners surveyed wanting to implement the Goals to various degrees. Lack of local indicators, tools and methods to support the implementation of SDGs in construction; knowledge regarding the SDGs; and costs associated with implementation were the main barriers identified. Therefore, to address these barriers, the development of new tools and methods to assist in practical application, and education amongst all the actors in the supply chain, was identified. Professional industry organisations need to lead the way to support industry adoption of the SDGs. The SDGs need to be involved in early goal setting, rather than being brought later into the project. Research by Kummen et al. (2023) focused on the need to reuse building materials for supporting circularity and achieving the aims of the SDGs. Through their study of Norway, the authors demonstrate that time, costs, lack of logistics, and requirements relating to materials, construction products and building design are the main barriers, especially when compared to other northern European countries. It is critical that there are structural changes to how reuse takes place, and education needs to support changes to praxis.

Tools

Tools to interlink with SDGs are needed. Green building rating systems generally have a very narrow sustainability scope. Goubran et al. (2023) assess how green, sustainable building and real estate standards have influenced the achievement of the SDGs. They found that less than 20 per cent of the green and sustainable building and real estate standards have impacted the SDGs. They add that fewer than 10 per cent of the standards are focused on the transformative change needed by the SDGs. They fear that there may be a lot of greenwashing instead, and that the claims made of achieving the SDGs by tools and benchmarks may actually be overstated as empirical evidence is lacking. Better linking and engagement with the standards and green building tools are critical.

The role of life cycle assessments (LCAs) in understanding the impacts of buildings, especially the environmental impact, is critical. When linked with the SDGs, they can assist with clear decision making for sustainability at the various levels, including the process, product, organisational or even higher level(s). To date, however, there are synergies between LCA and the SDGs but the two frameworks are not integrated (Cordella et al. 2023). In particular, LCSA (life cycle sustainability assessment) may be used to ensure products' improved sustainability. Most scholars have focused on the qualitative relationship between SDGs and LCSA; however, few authors have focused on a quantitative approach. There is no accepted, harmonised way, and this is yet to be further developed. Working on harmonising the SDGs beyond the goals, to the target and indicator levels, adds a bit of complexity as there are 169 targets and 231 unique indicators. The SDGs focus on socio-economic indicators, whereas the LCSA focuses on the environmental indicators, which is where LCAs have originated.

Research by Patyal et al. (2022) involves a study mapping the links between Industry 4.0 technologies, CE for sustainable operations and the SDGs. The systematic literature review showed that technologies assist with sustainable operations and in practices of CE. Selected SDGs these are mapped against are SDGs 6, 7, 9, 12 and 13. In a study by Goubran (2019), SDGs related directly and indirectly to the construction sector and real estate were examined. It was found that 17 per cent of the SDGs are directly dependent and 27 per cent indirectly dependent on building, construction and real estate activities. The largest contributions were attributed to SDGs 11, 6 and 7. Using content analysis, the research revealed that 44 per cent of the agenda's targets spanning all 17 SDGs are inherently connected to construction and real estate activities. Six of the 17 SDGs had a direct dependence on construction and real estate activities and 12 of the 17 had at least one target directly dependent on this sector. Over 80 per cent of the targets across SDGs 11, 6 and 7 are dependent on construction and real estate activities. Most (86 per cent) of the environmentally focused outcomes were also dependent on this sector. In another paper, by Scrucca et al. (2023), the author discusses how the selection of a comprehensive set of indicators and their linkages to the various SDGs needs to be quantitatively calculated, using LCAs, based on the specific representative value of each indicator referring to the building area and the statistical data on the yearly built area.

Krauss et al. (2022) show that economy and ecology interconnections across SDGs 8, 9, 12, 13 and 15 are focusing on universality, absences and modernity. In their study, they investigated the equity implications of these framings using indicator data to illustrate inherent injustices and inequities. In meeting the aims of the SDGs, recognising that trade-offs and equity shortcomings are key prerequisites to attaining genuine transformations for justice and sustainability through the SDGs also needs consideration.

Johnsson et al. (2020) state that SDGs may be used for decarbonising a business, but this should be driven by both supply and value chain perspectives, covering scopes 1, 2 and 3. There should be clear targets on climate change mitigation in line with the Paris Agreement and the main SDGs responding to climate change mitigation measures. However, it is not helpful if singular SDGs are cherry-picked, and there needs to be an approach of holistic assessment. Research undertaken by Fei et al. (2021) showed that the construction industry has a critical role in achieving the 17 SDGs. Using surveys and semi-structured interviews, 10 key SDGs were highlighted as being relevant for the industry: in order of importance, SDGs 11, 13, 6, 12, 9, 15, 5, 3, 7 and 8.

The next section examines studies on building materials and the SDGs.

Building Materials

In a study by Sharma et al. (2022), the use of phase-change materials (PCMs) and the use of nanomaterials and other cementitious materials have been mapped against the UN SDGs, particularly SDGs 7, 11 and 12. The authors argue that such new types of approaches will support the ultimate goal of attaining net zero carbon emissions from building infrastructure for a sustainable future. Sand, gravel and crushed stone are the building blocks of infrastructure, buildings and construction. Sustainable development is not possible with the mining of the aforementioned materials as it has a negative impact on the environment and humankind as a whole. The effect on health through direct and indirect impacts, system-level impacts on the ecosystem and the services they provide, and meeting projected targets into the future create tensions with improving the quality of life for all (Bendixen et al. 2021).

Patel and Ruparathna (2023) compare the life cycle thinking for road-based infrastructure evaluation, comparing the life cycle sustainability performance of alternative road pavement construction methods using building-information-modelling-based visual programmes. They demonstrate that the life cycle sustainable assessment framework assists infrastructure managers in selecting the most sustainable road construction method, in turn supporting the SDGs.

Roy et al. (2023) discuss the importance of the SDGs and waste management. They state that cooperation is critical in carrying out waste management practices, and suggest various theoretical models for the same. Effective policies and regulations need to be developed, and innovation and investment in waste management policies need attention. Waste management and its nexus with the built environment should be seen as a critical issue urgently. In this context, Ahmed (2023) discusses the opportunities arising from the use of plastics in construction, and its advantages, limitations and different applications. The study discusses the environmental and economic feasibility of using plastics in construction, including in the local construction sector. Economically, plastics can be used in the construction industry with the advantages of not ending up in landfill or as environmental pollutants in land and waterways if collection systems are in place. Plastics have good insulation properties, so they can be used instead of standard insulation, and can also create jobs through developing a collection process system. Environmental benefits are reduced pollution, mitigating climate change by reusing materials already available and reducing the use of construction materials.

What is very clear from this literature review is that while the industry is still struggling to understand the relationship of the SDGs with the industry, it is also clear that CE is not being considered in this context. Fragmented research exists and not all is linked with the SDGs. Hence, the SBC programme has identified the need to undertake research examining how CE links with the SDGs. Despite the criticisms of the SDGs not focusing on cultural issues and indigeneity, they still offer a global canvas to support the work of the SBC programme, particularly considering the timeframe of the said programme, from the initial timeline of 2022 to 2030, to meet the SDGs' timeframe.

The SDG Mapping Process and Attendant Survey

Nowhere in the 169 targets or the associated indicators is there any mention of CE. Despite the apparent lack of connection between the SDGs and CE, the principles underpinning sustainability and CE are the same: the quest to equally consider social, economic and environmental imperatives. The circularity aspects in the built environment are well covered in SDGs 8, 9, 11, 12 and 13, and more specifically addressed in SDGs 3, 4, 6 and 7.

As outlined in the *Transforming Our World: The 2030 Agenda for Sustainable Development* (UN 2023b), the SDGs, with targets and associated indicators, are integrated with each other. They balance the three dimensions of sustainable development across environmental, economic and social considerations. The 2030 Agenda focuses on *People*, to end poverty and hunger, and ensure they live in dignity and equality in a healthy environment. It focuses on *Planet*, to ensure that the planet is protected from degradation for the present and future generations. It focuses on *Prosperity*, to ensure that all citizens on the planet lead prosperous lives in harmony with nature. It focuses on *Peace*, to cultivate peaceful, just and inclusive societies. It focuses on *Partnerships*, to ensure that we collaborate to ensure that the aims of the SDGs are realised.

CE is also about ensuring that social and technological progress continues without adverse environmental impacts. While there are many definitions of CE, it is about designing out waste, keeping materials and products circulating in the system, optimising resources, providing long-lasting solutions so we do not draw on the use of virgin resources and regenerating natural systems. It has also been corroborated in other documents, such as Habitat III, the UNEA4 SCP innovative sustainable and consumption practices (UN 2019a, 2019b) and the SDG progress reports (ECOSOC 2018, 2019; UN 2021, 2022, 2023a).

The SBC programme started planning for the development of a scalable assessment framework for the state of CE of the buildings and construction sector at the national level in 2018. There were three reasons to undertake this understanding of circular built environments and to start developing a survey instrument. The first was due to the existence of the SDGs and its timeframe to 2030. The 15-year timeframe and, in particular, the 10-year horizon left for the application of the SDGs made logical sense when this study was undertaken. The second reason was the development of a reporting scheme for the Paris Agreement at COP 21, where countries were planning their Nationally Determined Contributions (NDCs) and the development of National Adaptation Plans (NAPs) (UNFCCC 2023) (resolution at COP 16, guidelines adopted at COP 17), and links to the Sendai framework. As stated in Chapter 1, the role of the built environment in GHG emissions results in high impacts, particularly in the developing regions of the world. Therefore, this research work is underpinned by an urgent need to ensure that building and construction practices are empirically aligned to the SDGs, and that the SDGs also support the principles of the CE. It is essential to 'lock in' circular practices of planning, design operation and end of life, rather than the current linear trajectory that the developed world has followed to date. The third reason is the fragmented nature of the tools and metrics, as demonstrated in 'Progress on the SDGs' above. Rather than creating something new, the intent was to develop the existing frameworks already available and link these to create a more holistic approach to circular built environments.

This study of the SDGs can be considered to be an explorative study for understanding the drivers and barriers for circularity, both at the sustainable development goal level and subsequently at the target and indicator levels. The intent in undertaking this work is not to seek confidence from the survey instrument, but rather to use the responses as an investigative starting point for further discussion in mapping the built environment from a circularity perspective with the SDGs. The targets and indicators are provided by the UN, but as these have been reviewed over a period of time, the exact wording may have changed since the deployment of the survey. It must also be noted that yearly progress reviews of the SDGs have brought changes in fine-tuning the targets and indicators, so while nuances may change, the essential intents underlining the targets and indicators are essentially consistent.

The state-of-play reports were produced by the SBC programme, as indicated in the previous chapter (Iyer-Raniga and Huovila 2020). Based on the literature and specifically that covering the SDGs, from the total of 248 indicators of the SDGs, 58 (UN 2023b) were identified by the SBC programme team for respondents to choose from for a survey. The survey was deployed from May 2020 to determine if these indicators resonated with the SBC programme's network partners and experts. It was kept open for a year beyond the timeline of the SBC programme's launching of these reports. The survey undertaken was done at two levels: at the level of the overall goal and then using a more detailed set of indicators across the various targets and indicators. These are presented in more detail with attendant analysis in the following section. The indicators were then examined against each of the regional reports to understand, in order of importance, first- and second-level indicators to determine the urgency in these regions to chart a way forward. The survey respondents were from various regions but were skewed to developed countries. While this gave an overall signal of which are the indicators particular to the built environment, it did not provide an understanding of the regional variations in climate, building and construction materials used, skills, use of technology, digitalisation, and so on.

To counteract this and to ensure capture of regional nuances, in addition to the global survey, workshops across Africa, Asia and Latin America were undertaken with local experts prior to the launch of the state-of-play reports undertaken by the authors for the global report. Ethics approval was obtained. The workshops were undertaken at convenient time zones for each of the participants. Each workshop lasted at least an hour, with some going on for up to 75 minutes. Workshops followed a consistent format where the authors of the report presented the background and the context, followed by regional author presentations. Then, themed discussions took place with online questions and real-time responses that supported rich discussions in the regions and, where appropriate, across regions. The overall numbers of workshop participants in each region were not high, due to issues associated with the pandemic; nevertheless, productive discussions resulted.

Three virtual circular built environment workshops were organised in 2020: on 15 October in Africa, 19 October in Latin America and 28 October in Asia. The participants were first introduced to the main findings from the *State of Play for Circular Built Environment* reports in Africa (Gibberd 2020), Latin America (Moreno 2020) and Asia (Niazi et al. 2020). The Oceania state-of-play report was used to compare the findings of the survey for the Australian respondents (Iyer-Raniga 2020). The state-of-play reports prepared for Europe (Westerholm 2020), Gulf Cooperation Council (GCC) countries (Al-Alawi et al. 2020) and North America (Keena and Dyson 2020) were not workshopped further. The SDGs and indicator survey results were shared, discussed and prioritised in each regional workshop. Since the number of responses from each region remained rather low, no direct conclusions could be drawn from the survey results. However, they formed the basis for facilitated discussions in each of the regional workshops and provided regional insights.

Some of the results were presented in papers (Iyer-Raniga and Huovila 2022) focusing on the Global South. This chapter presents the findings from the full survey (N = 185). The Appendix provides a list of all the primary and secondary indicators for each of the regions with detailed information as per the SDG indicators available.

MAPPING RESULTS AND ANALYSIS

Complete responses from the survey over the 2020–2022 period were further analysed to understand the priorities for different regions and variations in the SDG indicators. The total number of responses was 185 (N = 185). Thus, the overall survey response is denoted by N (N = 185) and the regional surveys are denoted by n (n = selected region).

As indicated in the previous section, the analysis was undertaken at two levels: the SDG level and the indicator level. Each of these was then further analysed based on the regional variations; it made logical sense to group these individually. At the SDG level, since the numbers of responses from Asia, Africa and Latin America was a smaller proportion of the overall response, it was reasonable to assess this at an overall goal level only. Exploratory analysis at the indicator level was undertaken across the individual regions, and this was supplemented by focus group workshops to validate the survey results as mentioned.

Goal Level

At the overall goal level, it was not surprising to see that the SDGs related to the built environment were rated highly by the participants. The two levels of analysis took place at the overall global level and then across the Global North (comprising the EU and Australia) and the Global South (comprising Africa, Asia and Latin America). The responses are presented in Table 5.1.

Table 5.1   SDG responses at goal level: global

Sustainable Development Goals	All [N = 185]
SDG 11 Resilient and Sustainable Cities	78%
SDG 12 Sustainable Consumption and Production	76%
SDG 13 Climate Change	72%
SDG 9 Sustainable Industrialization	63%
SDG 7 Access to Energy	50%
SDG 6 Water and Sanitation	46%
SDG 8 Economic Growth and Productive Employment	45%
SDG 3 Health and Well-being	36%
SDG 17 Global Partnerships	34%

SDG 11 targets sustainable cities, while SDG 12 is all about responsible consumption and production. The importance of climate change has been highlighted as being more important than industrialisation, with energy, water and economic growth coming at the lower levels of ranking. Health and partnerships represented just a third of the responses; clearly, they do not directly impact the built environment.

Table 5.2   SDG responses at the goal level: global and Global North

Sustainable Development Goals	All [N = 185]	North [n = 88] (EU n = 57, AU n = 17, North America n = 10, Middle East n = 4)
SDG 11 Resilient and Sustainable Cities	78%	84%
SDG 12 Sustainable Consumption and Production	76%	80%
SDG 13 Climate Change	72%	78%
SDG 9 Sustainable Industrialization	63%	65%
SDG 8 Economic Growth and Productive Employment	45%	48%
SDG 7 Access to Energy	50%	45%
SDG 6 Water and Sanitation	46%	36%
SDG 17 Global Partnerships	34%	36%
SDG 3 Health and Well-being	36%	32%

For the Global North, the results are aligned to that of the global responses, except for SDG 8 on economic growth, which was ranked more highly, as well as SDG 17 on partnerships (see Table 5.2).

Table 5.3   *SDG responses at the goal level: global and Global South*

Sustainable Development Goals	All [N = 185]	South [n = 97] (Africa n = 39, Asia n = 35, LAC n = 23)
SDG 12 Sustainable Consumption and Production	76%	73%
SDG 11 Resilient and Sustainable Cities	78%	72%
SDG 13 Climate Change	72%	67%
SDG 9 Sustainable Industrialization	63%	62%
SDG 7 Access to Energy	50%	55%
SDG 6 Water and Sanitation	46%	55%
SDG 8 Economic Growth and Productive Employment	45%	42%
SDG 3 Health and Well-being	36%	40%
SDG 17 Global Partnerships	34%	32%

When comparing the global set of responses with the Global South, SDG 12 on responsible sustainable consumption and production ranked the highest for the latter, instead of SDG 11 on resilient and sustainable cities. The rest followed the same rankings in importance as that of the global (see Table 5.3).

Indicator-Level Responses

For all responses, the core indicators were selected across the 240-plus indicators available. The overall response rate above 50 per cent of the total responses was analysed, shown in Table 5.4 based on the levels of importance arising from the surveys. Note that where the text appears as bold, this refers to the core indicators, and where italicised, secondary indicators. As stated earlier, the core indicators were derived by the SBC programme, regarding what SDG indicators were critical for achieving circular built environments. Where the text is neither bold nor italicised, this indicator reflects the responses from the survey but is not considered as a primary or secondary indicator by the SBC programme team. Table 5.5 presents the verbatim text for each of the indicators identified in Table 5.4.

Global

Table 5.4 provides the primary and secondary indicators arising from the global survey with five primary or core indicators and nine secondary indicators. The indicator 11.7.1 just meets the 50 per cent cut-off point of importance in the ranking, but this was not included as part of the secondary indicators. As noted above, Table 5.5 provides descriptions of the indicators.

Table 5.4   *Five primary and nine secondary indicators arising from the global survey*

12.5.1 recycling rate	**80%**
8.4.1 material footprint	**79%**
12.2.1 material footprint	**79%**
11.c.1 local materials*	**76%**
9.4.1 CO$_2$ emission	**66%**
11.6.1 municipal solid waste	*60%*
13.2.1 NDCs and NAPs	*60%*
6.3.1 wastewater treated	*59%*
6.4.1 water-use efficiency	*58%*
7.2.1 renewable energy	*57%*
11.1.1 informal settlements	*57%*
7.1.2 reliance on clean fuels	*55%*
12.a.1 environmentally sound technologies*	*54%*
12.7.1 public procurement	*54%*
11.7.1 green and public spaces: share of the built-up area	50%**

Note: * Changed from original wording of targets and/or indicators subsequent to the research undertaken. ** 11.7.1 at 50% was not chosen as the secondary indicator; only the indicators in italic were chosen.

Table 5.5   *Verbatim text from the UN website of the five primary*
indicators and nine secondary indicators arising from the
global survey[1] (N = 185)

Primary indicator	Secondary indicator
12.5.1 National recycling rate, tons of material recycled	11.6.1 Proportion of municipal solid waste collected and managed in controlled facilities out of total municipal waste generated by cities
8.4.1 Material footprint, material footprint per capita, and material footprint per GDP	13.2.1 Number of countries with nationally determined contributions, long-term strategies, national adaptation plans and adaptation communications, as reported to the secretariat of the United Nations Framework
12.2.1 Material footprint, material footprint per capita, and material footprint per GDP	6.3.1 Proportion of domestic and industrial wastewater flows safely treated
11.c.1 Proportion of financial support to the least developed countries that is allocated to the construction and retrofitting of sustainable, resilient and resource-efficient buildings utilizing local materials; a suitable replacement indicator is under development to be proposed for the 2025 comprehensive review*	6.4.1 Change in water-use efficiency over time
9.4.1 CO_2 emission per unit of value added	7.2.1 Renewable energy share in the total final energy consumption
	11.1.1 Proportion of urban population living in slums, informal settlements or inadequate housing
	7.1.2 Proportion of population with primary reliance on clean fuels and technology
	12.a.1 Installed renewable energy-generating capacity in developing countries (in watts per capita); updated from Amount of support to developing countries on research and development for sustainable consumption and production and environmentally sound technologies in the earlier version*

Primary indicator	Secondary indicator
	12.7.1 Degree of sustainable public procurement policies and action plan implementation
	**11.7.1 Average share of the built-up area of cities that is open space for public use for all, by sex, age and persons with disabilities

Note: * Changed from original wording of targets and/or indicators subsequent to the research undertaken. ** 11.7.1 at 50% was not chosen as a secondary indicator. However, for the individual regions any indicator above 50% was considered.

The survey responses at the target/indicator level were then analysed for various regional variations. The objective was to determine how relevant these indicator sets are in different regions to assess their applicability, set objectives and monitor trends in the built environment. Where the indicators had the same ranking, they are organised in order of the SDG rankings. For the overall global rankings, the highest ranked was 12.5.1 recycling rate (80 per cent) and 9.4.1 CO_2 emissions (66 per cent) and the secondary indicators commenced with 11.6.1 municipal solid waste (60 per cent) and ended with the indicator dealing with open space for public use (11.7.1) at 50 per cent.

For the Global North (EU and Australia), there were very clear groupings in the overall responses, but with the Global South the various indicators were more spread out. For example, the most important indicator for the global response (N = 185) was 12.5.1 recycling rates (80 per cent), followed by 8.4.1 material footprint/12.2.1 material footprint (79 per cent), 11.c.1 local materials (76 per cent) and 9.4.1 CO_2 emissions (66 per cent). The secondary indicators ranged from 11.6.1 municipal solid waste to 11.7.1 open space for public use (60 per cent to 50 per cent drop in importance). Between these were 13.2.1 NDCs and NAPs, 6.3.1 wastewater treated, 6.4.1 water-use efficiency, 7.2.1 renewable energy, 11.1.1 informal settlements, 7.1.2 reliance on clean fuels, 12.a.1 environmentally sound technologies and 12.7.1 public procurement. For the same rankings for Australia, the percentages of importance were 71 per cent for primary indicator 8.4.1. material footprint, and for the secondary indicators from 59 per cent to 53 per cent in descending order from local material use (11.c.1) to marine pollution (14.1).

Europe

For the EU region, the primary indicators all matched the overall responses to the survey (N = 185). The EU was the most dominant responder at n = 57, roughly a third of the overall numbers. The responses for the EU are shown in Table 5.6.

Table 5.6   Primary and secondary indicators for the EU

8.4.1 material footprint	**93%**
12.2.1 material footprint	**93%**
12.5.1 recycling rate	**89%**
11.c.1 local materials*	**75%**
9.4.1 CO$_2$ emission	**69%**
13.2.1 NDCs and NAPs	*56%*
11.6.1 municipal solid waste	*55%*
7.2.1 renewable energy	*55%*
11.1.1 informal settlements	*54%*
*12.a.1 environmentally sound technologies**	*50%*

Note: * Changed from original wording of targets and/or indicators subsequent to the research undertaken.

The EU follows very closely to the global survey, as its response is roughly a third of the total. The only differences in the responses lie with the secondary indicators, in not having the links with water. Indicators 6.3.1 on wastewater and 6.4.1 on water efficiency, both of which appear in the global survey as secondary indicators, are absent. Furthermore, indicator 7.1.2 on clean fuels and 12.7.1 on public procurement were also not picked in the EU set of indicators. There may be several reasons for this. First is that the EU is a conglomeration of countries with climate variations, so the importance of water as a resource may vary from country to country. Clean fuels are already being rolled out in the EU, so this may not be seen as a core indicator for a circular built environment as it is considered mainstream already. Public procurement is also part of the policy context of the EU, so public procurement may not be seen as essential in the EU context.

Australia

The total number of responses in Australia was very low at 17. This represented less than 10 per cent of the overall responses. Yet, the responses are reflective of the relationship between key issues facing the country and the role of its built environment, as shown in Table 5.7.

Table 5.7   Primary and secondary indicators for Australia

8.4.1 material footprint	71%
12.5.1 recycling rate	71%
12.2.1 material footprint	71%
11.6.1 municipal solid waste	*65%*
11.c.1 local materials*	59%
6.4.1 water-use efficiency	*59%*
7.1.2 reliance on clean fuels	*59%*
12.7.1 public procurement	*59%*
9.4.1 CO_2 emission	53%
11.7.1 green and public spaces: share of the built-up area	53%
7.3.1 energy efficiency: energy intensity in terms of primary energy and GDP	53%
11.1.1 informal settlements	*53%*
14.1.1 marine pollution: coastal eutrophication and plastic debris density*	53%

Note: * Changed from original wording of targets and/or indicators subsequent to the research undertaken.

Findings from Australia follow a similar trajectory to those from the EU. However, the percentages are slightly different in the Australian context, with the highest ranking at 71 per cent for 8.4.1 material footprint and the lowest 53 per cent for 14.1.1 marine pollution, across both the primary and secondary indicators. For Australia, amongst the key primary indicators is 11.6.1 (municipal solid waste) at 65 per cent, which has come up as a secondary indicator in the global survey. The reason for this is because of the China Sword Policy[2] in Australia; municipal solid waste was an important issue (and still is) as the waste had to be managed within the country (Iyer-Raniga 2020). The global primary indicator 9.4.1 on CO_2 emissions becomes a secondary indicator in the Australian survey responses. Also, other topical issues in the country are highlighted, such as indicator 11.1.1 of inadequate housing and target 14.1 on reducing marine pollution showing up as secondary indicators.

Africa

The total number of responses from Africa was 39. Compared to the total, this represented only a fraction above 21 per cent. However, the responses indicate the intricacies of the continent, with its spread of economies, climate, peoples and culture, as shown in Table 5.8.

Table 5.8   Primary and secondary indicators for Africa

12.5.1 recycling rate	**86%**
11.c.1 local materials*	**82%**
6.3.1 wastewater treated	*77%*
6.4.1 water-use efficiency	*74%*
13.2.1 NDCs and NAPs	*74%*
12.2.1 material footprint	**74%**
8.4.1 material footprint	**73%**
9.4.1 CO$_2$ emission	**73%**
11.6.1 municipal solid waste	*73%*
11.1.1 informal settlements	*71%*
7.2.1 renewable energy	*62%*
7.b.1 infrastructure and technology: investments in energy efficiency as a % of GDP*	62%
13.3.1 awareness-raising: countries with impact reduction and early warning, curricula	62%
11.7.1 green and public spaces: share of the built-up area	59%
11.a.1 resource efficiency policies: local governments implementing risk reduction strategies*	58%
12.a.1 environmentally sound technologies*	*58%*
7.3.1 energy efficiency: energy intensity in terms of primary energy and GDP	56%
11.b.2 resource efficiency policies: countries with national and local risk reduction strategies	55%
7.1.2 reliance on clean fuels	*54%*
7.a.1 clean energy research and technology: mobilized amount of USD*	54%
12.8.1 sustainable development awareness: extent to education mainstreamed	53%
12.7.1 public procurement	*53%*
1.4.1 rights to economic resources: access to basic services	51%
2.4.1 food production: sustainable agriculture	51%

Note: * Changed from original wording of targets and/or indicators subsequent to the research undertaken.

Africa's challenges are quite distinct and spread over a number of SDG indicators, as can be observed from Table 5.8. Materials, water, CO$_2$ emissions, housing, energy use, education and resource efficiency are all highlighted. The top six indicators are a combination of the global set of primary and secondary indicators, with three each of the global indicators making it to the top six. High on the rankings are other global primary and secondary indicators, being 8.4.1 material footprints (73 per cent), 9.4.1 CO_2 emissions (73 per cent), 11.6.1 municipal solid waste (73 per cent) and 11.1.1 informal settlements (71 per cent). Four secondary indicators from the global list make it into the remaining secondary indicators for Africa: 7.2.1 renewable energy, 12.a.1 environmentally sound technologies, 7.1.2 reliance on clean fuels and 12.7.1 public procurement. The other indicators are a mix of investments in energy and energy efficiency, education, planning and local government policies, including those for disaster risks, food and basic services. Indicator 1.4.1 appears in both Africa and Latin America and indicator 2.4.1 appears only in the African responses. This spread highlights the complexity of the challenges in Africa.

When engaging with the experts in Africa, the workshop participants prioritised indicator 7.3.1 on energy intensity, and indicator 13.3.1 awareness-raising: countries with impact reduction and early warning systems and curricula. In addition, 12.5.1 on recycling was highlighted as being important, followed by 6.3.1 on wastewater treatment and indicator 11.6.1 on municipal solid waste. Material footprint (indicator 8.4.1/12.2.1) followed next. Then, a second set of high-scoring indicators were prioritised where indicator 9.4.1 on CO_2 emission topped the list with indicator 11.1.1 on inadequate housing and 11.c.1 on local materials following closely behind. Indicator 7.2.1 on renewable energy and indicator 12.a.1 on environmentally sound technologies or indicator 12.a.1 on renewable energy were not found to be important. A similar exercise was then carried out on the third set of indicators, where public procurement, expressed in indicator 12.7.1, was found to be most important. The following quote from a workshop participant captures the importance of procurement: 'For procurement, just relying on government is not enough, there is a need to bring in private actors.'

Asia

The responses from Asia numbered 35 (n = 35), slightly lower than that of Africa. These responses are not demonstrative of the same spread of Africa; however, they do provide some useful insights.

Table 5.9   Primary and secondary indicators for Asia

11.c.1 local materials*	**77%**
8.4.1 material footprint	**69%**
12.2.1 material footprint	**69%**
12.5.1 recycling rate	**60%**
6.4.1 water-use efficiency	*57%*
11.1.1 informal settlements	*57%*
12.7.1 public procurement	*57%*
13.2.1 NDCs and NAPs	*57%*
11.b.1 national risk reduction strategies	54%
11.7.1 green and public spaces: share of the built-up area	54%
12.1.1 10YFP on SCP: countries with national action plan	54%
17.18.1 SDG monitoring	54%
6.3.1 wastewater treated	*51%*
7.1.2 reliance on clean fuels	*51%*
11.6.1 municipal solid waste	*51%*
*12.a.1 environmentally sound technologies**	*51%*

Note: * Changed from original wording of targets and/or indicators subsequent to the research undertaken.

Asia's challenges are slightly different from Africa, as shown in Table 5.9. Indicator 11.c.1 local materials use was the most highly ranked indicator, followed by indicator 8.4.1/2.2.1 material footprint and 12.5.1 on recycling rates. The remaining secondary indicators appeared the same as in the global survey except for considerations on 11.b.1 national risk reduction strategies, 11.7.1 green and public spaces, 12.1.1 countries with national action plans and 17.18.1 SDG monitoring. Lower on the rankings were 6.3.1 wastewater treatment, 7.1.2 reliance on clean fuels, 11.6.1 municipal solid waste and 12.a.1 environmentally sound technologies.

At the Asia workshop, the participants selected indicator 11.7.1 'Average share of the built-up area of cities that is open space for public use for all, by sex, age and persons with disabilities' and indicator 12.1.1 'Number of countries developing, adopting or implementing policy instruments aimed at supporting the shift to sustainable consumption and production'. In addition, the workshop participants prioritised first potential indicators, finding material footprint (8.4.1) and local materials (11.c.1) to be the most important, followed by inadequate housing (11.1.1) and open space for public use (11.7.1). Prioritisation of a second set of indicators ranked policy instruments supporting the shift to sustainable consumption and production (12.1.1), recycling (12.5.1), public procurement (12.7.1) and nationally determined contributions (13.2.1) before wastewater treatment (6.3.1).

A similar exercise was then undertaken on the third set of indicators where 11.b.1 'Number of countries that adopt and implement national disaster risk reduction strategies in line with the Sendai Framework for Disaster Risk Reduction 2015–2030' appeared before environmentally sound technologies or renewable energy (12.a.1) and CO_2 emission (9.4.1), followed by 11.3.2 'Proportion of cities with a direct participation structure of civil society in urban planning and management' that operate regularly and democratically. Water-use efficiency (6.4.2) and clean fuels (7.1.2) were not seen as that important by local experts. The discussion focused on a lack of engagement with materials, as one participant highlighted: 'In a market economy it is very difficult to single out building materials. When you go into practice, it is all about what clients are demanding … the scale of engagement is important.'

Latin America and the Caribbean (LAC)
The responses attributed to this region represented only 12.4 per cent of the total responses at 23 (Table 5.10), slightly higher than Australia. As in the case of Africa, the responses are quite spread, even more than those of Africa; particularly the secondary indicators. While there is alignment with the top core indicators and some alignment with secondary indicators, 12.8.1 on education, 11.b.1 risk reduction strategies at national and local levels (which also appears in Asia), 13.3.1 awareness-raising and curricular development, 9.b.1 domestic technology development, 7.a.1 clean energy research, 1.4.1 access to basic services (as in the case of Africa), 11.3.1 citizen participation, 5.5.2 gender issues, 9.5.1 using research and development expenditure as a percentage of GDP, and 15.1.2 and 15.a.1 protecting ecosystems and biodiversity with the use of overseas aid, were raised as critical issues from the survey.

Table 5.10   *Primary and secondary indicators for LAC*

12.5.1 recycling rate	**83%**
12.2.1 material footprint	**78%**
8.4.1 material footprint	**78%**
9.4.1 CO$_2$ emission	**78%**
11.c.1 local materials*	**78%**
*12.a.1 environmentally sound technologies**	*73%*
6.3.1 wastewater treated	*70%*
11.1.1 informal settlements	*70%*
12.8.1 sustainable development awareness: extent to education mainstreamed	70%
7.1.2 reliance on clean fuels	*68%*
7.2.1 renewable energy	*65%*
11.b.1 resource efficiency policies: local governments implementing risk reduction strategies	65%
12.7.1 public procurement	*65%*
6.4.1 water-use efficiency	*61%*
13.2.1 NDCs and NAPs	*61%*
13.3.1 awareness-raising: countries with impact reduction and early warning, curricula	61%
9.b.1 domestic technology development: proportion of medium and high-tech industry value added	57%
11.6.1 municipal solid waste	*57%*
11.b.2 resource efficiency policies: countries with national and local risk reduction strategies	57%
7.a.1 clean energy research and technology: mobilized amount of USD*	55%
1.4.1 rights to economic resources: access to basic services	52%
5.5.2 women's participation: proportion of women in managerial positions	52%
8.3.1 policies supporting productive activities: proportion of informal employment	52%
9.5.1 technological capabilities: research and development expenditure as a proportion of GDP	52%
11.3.1 inclusive urbanization: proportion of cities with a direct participation structure*	52%

15.1.2 ecosystems: important sites for terrestrial and freshwater biodiversity	52%
15.a.1 biodiversity: ODA and public expenditure on conservation and sustainable use	52%

Note: * Changed from original wording of targets and/or indicators subsequent to the research undertaken.

For Latin America, workshop participants prioritised the following: 4.7.1/12.8.1 'Extent to which (i) global citizenship education and (ii) education for sustainable development are mainstreamed in (a) national education policies; (b) curricula; (c) teacher education; and (d) student assessment' and 11.b.1 'Number of countries that adopt and implement national disaster risk reduction strategies in line with the Sendai Framework for Disaster Risk Reduction 2015–2030'. These were followed by inadequate housing (11.1.1) and local materials (11.c.1), with material footprint (8.4.1 and 12.2.1) and renewable energy (7.2.1) coming last.

The points raised in the workshop were that in rural environments, reusing and recycling materials is common; however, buildings are not generally designed for deconstruction. Data is generally a problem, and awareness of reuse and recycling of building materials is low. Traditionally, recycling of bottles, paper and metals is quite common. The informal sector focuses on materials for immediate use, such as scrap iron or glass. Typically, buildings are just demolished rather than deconstructed to enable a second (or more) life of materials. Recycled materials have a stigma, generally associated with low socio-economic parts of the society. New materials are to be used only in new construction. This was highlighted by a workshop participant who said: 'In your new house you don't want to use old windows or old tiles, you want everything new.'

A second set of high-scoring indicators were prioritised, where education (4.7.1 and 12.8.1) was seen most important, followed by recycling (12.5.1) and public procurement (12.7.1). Indicators for CO_2 emission (9.4.1) and municipal solid waste (11.6.1) were not seen as important by this expert group. The importance of education was raised several times. A challenge lies in how to blend traditional knowledge with modern times and how to inculcate understanding of natural cycles to people who live in a concrete environment remote from nature. A similar exercise was then undertaken on a third set of indicators, where in addition to wastewater treatment (6.3.1), water-use efficiency (6.4.1), clean fuels (7.1.2) and education (13.3.1, same as 4.7.1 and 12.8.1), 11.5.2 'Direct economic loss in relation to global GDP, damage to critical infrastructure and number of disruptions to basic services, attributed to disasters' was also seen to be important.

DISCUSSION

The GSDR reports have shown that advancement on the SDGs has stalled. Some SDGs have progressed, particularly in research and development funding, whereas the key threats to humanity through increased emissions and climate change have yet to be addressed through the relevant sustainable development goals, targets and indicators. The SDGs underpinning the planning, design, construction, operation and end of life of the built environment need more work. Housing is an immediate and significant need in both the developed and developing world.

As Sachs et al. (2023) highlight, sustainable cities need investment for the long term, otherwise it is not possible to get to the end point. Reducing global emissions, preventing biodiversity loss, and responsibly addressing our production and consumption practices are all critical. It is in this context that the circular built environments play an important role in ensuring 'lock in' does not take place and the future of the built environment is planned in a very different way compared to the journey taken to date. As discussed in 'The Built Environment and the SDGs' above, research by Goubran et al. (2023) highlights that transformative change is needed, with fewer than 10 per cent of the standards considering the SDGs, building on the work that Goubran (2019) showed where 27 per cent of SDGs are indirectly associated with all building, construction and real estate. Gade and Opoku (2020) discussed building owners wanting to apply the SDGs but not necessarily having the knowledge of how this might happen. Bide et al. (2023) focus on the impact of materials use, now and into the future. The need to develop local indicators, tools and methods is important.

While the focus is on the process, social aspects of the built environment focusing on SDG 4 on education, SDG 5 on gender equality and SDG 10 on balancing the needs of countries is needed. Most case studies reported in the academic literature and especially those that link the studies with existing tools and assessment models for green building certification can be linked to SDGs 3, 6, 7, 9, 11, 12, 13 and 15. This is not surprising, because SDG 3 on health and well-being is linked to assessment schemes on thermal comfort; likewise, SDGs 6 and 7 on energy and water, SDG 9 on innovation, SDG 11 on cities and settlements, SDG 12 on consumption and efficiencies of various aspects of building construction and operation, SDG 13 on climate change and SDG 15 on biodiversity. Where low socio-economic groups are involved, SDG 1 on no poverty is also impacted. Attention needs to be provided to SDG 4 on education.

SDGs need to be considered early in the process of planning and design for the built environment. A holistic decision-making approach is needed where the SDGs, especially in developing countries, need to be linked to government policies, so that they are also linked to NDCs and NAPs. Structural changes need to take place across the board, with all the key factors that are drivers for the built environment but can also be challenges, in the form of time, costs, lack of logistics, materials, building design, construction methods, the use of technology and others, brought into the holistic approach. Education is needed, especially in the Global South, as highlighted in the regional workshops.

CONCLUSION

Ultimately, the principles underpinning circularity are the same as those of sustainability. The goals, targets and indicators associated with the SDGs in Agenda 2030 are a good starting point for achieving circularity in the built environment. While some of the indicators are specific to the year 2020, nevertheless they provide a good yardstick of measurement for monitoring and reporting progress. As the SDGs also support other related UN programmes such as the NDCs, alignment where possible supports concerted efforts towards achieving social, economic and environmental sustainability. From the exercise undertaken, while to some extent all SDGs are relevant, four are core for circular built environments. These are SDGs 9, 11, 12 and 13.

- Goal 9 concerns building resilient infrastructure and fostering industry and innovation. Circular economies need to reconsider their manufacturing processes and consider the small scale or local scale, as outlined in indicators 9.2.2 and 9.3.1. Emissions in the form of CO_2 or its equivalent due to industry are reflected in indicator 9.4.1. More investigations through research and development as a proportion of value added to GDPs are considered in indicators 9.5.1 and 9.b.1.
- Goal 11 is concerned with making cities and human settlements inclusive, safe, resilient and sustainable. Ensuring adequate housing, solid waste management, and use of local materials and technologies for construction and retrofitting is reflected in indicators 11.1.1, 11.6.1 and 11.c.1.
- Goal 12 focuses on consumption and production. Indicators 12.1 1, 12.2.1, 12.5.1 and 12.7 are critical indicators as they focus on sustainable consumption and production of materials and use, and footprints including recycling rates and per capita.
- Goal 13 on climate change and the urgent need to respond to and prepare for impacts on climate change is reflected in the form of proactive integration into national policies, strategies and planning. Indicators 13.2.1 on countries with NDCs, NAPs and long-term strategies, and 13.3.1 on education and curricula, are particularly being recognised as critical in the Global South.

The work undertaken for a circular built environment focused on a regional approach to the Global South, but this needs to be taken to the next step in considering the variations in people, climate, context, skills and know-how within the countries comprising the Global South. When considering the Global South and the case studies presented by the SBC programme, the following Goals are highlighted: SDGs 6, 7, 8, 9, 11, 12 and 17. The engagement at the national level with NDCs and NAPs is also critical. This exploratory work on the SDGs may become the launch pad for further engagement with the countries in the Global South that have the most need for and most urgency regarding circular built environments.

NOTES

1. The indicators are derived verbatim from the UN SDGs as per the indicators available (https://unstats.un.org/sdgs/indicators/indicators-list/).
2. In February 2017 China announced its National Sword initiative. Known as the China Sword Policy, this initiative bans the importation of various materials and products into China, effective from January 2018.

REFERENCES

Ahmed, N. (2023) Utilizing Plastic Waste in the Building and Construction Industry: A Pathway Towards the Circular Economy, *Construction and Building Materials*, 383, 131311. https://doi.org/10.1016/j.conbuildmat.2023.131311

Al-Alawi, B., Tibi, G., Cavallari, G., Cindric, H., Shaka, H., Gargallo, M., Hegazy, N., and Saqib, S. (2020) *State of Play for Circular Built Environment in Gulf Cooperation Council (GCC) Countries: Sustainable Building and Construction Programme Regional Report for GCC – Final Report October 2020*, ARUP and United Nations One Planet Network Sustainable Building and Construction Programme. https://www .oneplanetnetwork .org/ sites/ default/ files/ from -crm/ gcc _final_210214.pdf

Alfirević, N., Malešević Perović, L., and Mihaljević Kosor, M. (2023) Productivity and Impact of Sustainable Development Goals (SDGs)-Related Academic Research: A Bibliometric Analysis, *Sustainability*, 15, 7434. https:// doi .org/ 10 .3390/ su15097434

An, E. (2023) Accelerating Sustainability through Better Reporting Sustainability Accounting, *Management and Policy Journal*, 14(4), 904–914. https:// doi .org/ 10 .1108/SAMPJ-08-2022-0453

Bendixen, M., Iversen, L.L., Best, J., Franks, D.M., Hackney, C.R., Latrubesse, E.M., and Trusting, L.S. (2021) Sand, Gravel, and UN Sustainable Development Goals: Conflicts, Synergies, and Pathways Forward, *One Earth*, 4(8), 1095–1111. https:// doi.org/10.1016/j.oneear.2021.07.008

Bide, T., Novellino, A., Petavratzi, E., and Watson, C.S. (2023) A Bottom-Up Building Stock Quantification Methodology for Construction Minerals Using Earth Observation: The Case of Hanoi, *Cleaner Environmental Systems*, 8, 100–109. https://doi.org/10.1016/j.cesys.2023.100109

BRE (Building Research Establishment) (n.d.) *The Built Environment and Future Sustainability: The Relationship between BREEAM and the Sustainable Development Goals*, BREEAM. www.breeam.com

Chahdi, T.O., Bennouna, M., and Tahrouch, M. (2023) Towards Achieving Sustainable Development Goals in Academia: A Case of Good Practice, *African Scientific Journal*, 3(17), 260–295.

Cordella, M., Horn, R., Hong, S.H., Bianchi, M., Isasa, M., Harmens, R., Sonderegger, T., and Pikhola, H. (2023) Addressing Sustainable Development Goals in Life Cycle Sustainable Assessment: Synergies, Challenges and Needs, *Journal of Cleaner Production*, 415, 137719. https://doi.org/10.1016/j.jclepro.2023.137719

ECOSOC (Economic and Social Council of the UN) (2018) *E/2018/684 High-Level Political Forum on Sustainable Development, Report of the Secretary-General*, UN. https://sustainabledevelopment.un.org/content/documents/18541SG_SDG_Progress _Report_2018_ECOSOC.pdf

ECOSOC (Economic and Social Council of the UN) (2019) *Special Edition: Progress Towards the Sustainable Development Goals – Report of the Secretary-General*, UN. https://undocs.org/E/2019/68

Fei, W., Opoku, A., Agyekum, K., Oppon, J.A., Ahmed, V., Chen, C., and Lok, K.L. (2021) The Critical Role of the Construction Industry in Achieving the Sustainable Development Goals (SDGs): Delivering Projects for the Common Good, *Sustainability*, 13(16), 9112. https://doi.org/10.3390/su13169112

Gade, A.N., and Opoku, A. (2020) Challenges for Implementing the Sustainable Development Goals in the Danish Construction Industry: Building Owners' Perspective, *Proceedings of the 36th Annual ARCOM Conference*, UK, 7–8 September 2020, 615–624.

Gibberd, J. (2020) *State of Play for Circular Built Environment in Africa: Sustainable Building and Construction Programme Regional Report for Africa – Final Report May 2020*, Gauge and United Nations One Planet Network Sustainable Building and Construction Programme. https://www.oneplanetnetwork.org/sites/default/files/ from-crm/africa_final_201030.pdf

Goubran, S. (2019) On the Role of Construction in Achieving the SDGs, *Journal of Sustainability Research*, 1, e190020. https://doi.org/10.20900/jsr20190020

Goubran, S., Walker, T., Cucuzzella, C., and Schwarz, T. (2023) Green Building Standards and the United Nations' Sustainable Development Goals, *Journal of Environmental Management*, 326(PartA), 116552. https://doi.org/10.1016/j.jenvman .2022.116552

Hussain, K., He, Z., Ahmad, N., Iqbal, M., and Nazneen, S. (2023) Mapping Green, Lean, Six Sigma Enablers through the Lens of a Construction Sector: An Emerging Economy's Perspective, *Journal of Environmental Planning and Management*, 66(4), 779–812. https://doi.org/10.1080/09640568.2021.2006155

Ido, B.J., Macêdo, A.N., and Martins, V.W.B. (2023) Construction Industry and Its Contributions to Achieving the SDGs Proposed by the UN: An Analysis of Sustainable Practices, *Buildings*, 13(5), 1168. https://doi.org/10.3390/buildings13051168

Indana, F., and Pahlevi, R.W. (2023) A Bibliometric Approach to Sustainable Development Goals (SDGs) Systematic Analysis', *Cogent Business & Management*, 10(2), 2224174. https://doi.org/10.1080/23311975.2023.2224174

Iyer-Raniga, U. (2020) *State of Play for Circular Built Environment in Oceania: Sustainable Building and Construction Programme Regional Report for Oceania – Final Report April 2020*, RMIT University and United Nations One Planet Network

Sustainable Building and Construction Programme. https://www.oneplanetnetwork
.org/sites/default/files/from-crm/oceania_final_201101.pdf

Iyer-Raniga, U., and Huovila, P. (2020) *Global State of Play for Circular Built Environment.* https://www.oneplanetnetwork.org/sites/default/files/from-crm/global _final_210214.pdf

Iyer-Raniga, U., and Huovila, P. (2022) Mapping Sustainability Indicators for Circular Built Environment in the Global South, *IOP Conference Series: Earth and Environmental Science*, 1101, 062033. https://doi.org/10.1088/1755-1315/1101/6/ 062033

Johnsson, F., Karlsson, I., Rootzén, J., Ahlbäck, A., and Gustavsson, M. (2020) The Framing of a Sustainable Development Goals Assessment in Decarbonizing the Construction Industry – Avoiding 'Greenwashing', *Renewable and Sustainable Energy Reviews*, 131, 110029. https://doi.org/10.1016/j.rser.2020.110029

Kandpal, R., and Okitasari, M. (2023) Governance Transformation Towards Localisation of Sustainable Development Goal 11 in India, *World Development Sustainability*, 2, 100069. https://doi.org/10.1016/j.wds.2023.100069

Keena, N., and Dyson, A. (2020) *State of Play for Circular Built Environment in North America: Sustainable Building and Construction Programme Regional Report for North America – Final Report May 2020*, Yale CEA and United Nations One Planet Network Sustainable Building and Construction Programme. https://www .oneplanetnetwork.org/sites/default/files/north_america_final_201101.pdf

Krauss, J.E., Cisneros, A.J., and Requena-i-Mora, M. (2022) Mapping Sustainable Development Goals 8, 9, 12, 13 and 15 through a Decolonial Lens: Falling Short of 'Transforming Our World', *Sustainability Science*, 17, 1855–1872. https://doi.org/ 10.1007/s11625-022-01112-3

Kummen, T.M., Bohne, R.A., and Lohne, J. (2023) Mapping of Construction Materials Reuse Practices within Large Norwegian Municipalities, *IOP Conference Series: Earth and Environmental Science*, 1176, 012036. https://doi.org/10.1088/1755 -1315/1176/1/012036

Liu, Z., Zhang, Y., Ni, X., Dong, M., Zhu, J., Zhang, Q., and Wang, J. (2023) Climate Action May Reduce the Risk of Unemployment: An Insight into the City-Level Interconnections Among the Sustainable Development Goals, *Resources, Conservation & Recycling*, 1941, 07002. https://doi.org/10.1016/j.resconrec.2023 .107002

Moreno, P. (2020) *State of Play for Circular Built Environment in Latin America and the Caribbean (LAC): Sustainable Building and Construction Programme Regional Report for LAC – Final Report May 2020*, Ecosur Network and United Nations One Planet Network Sustainable Building and Construction Programme. https://www .oneplanetnetwork.org/sites/default/files/from-crm/lac_final_201101.pdf

Niazi, Z., Singh, A., and Sen, I. (2020) *State of Play for Circular Built Environment in Asia: Sustainable Building and Construction Programme Regional Report for Asia – Final Report April 2020*, TARA/Development Alternatives and United Nations One Planet Network Sustainable Building and Construction Programme. https://www .oneplanetnetwork.org/sites/default/files/from-crm/asia_final_201101.pdf

Norouzi, M., Chàfer, M., Cabeza, L., Jiménez, L., and Boer, D. (2021) Circular Economy in the Building and Construction Sector: A Scientific Evolution Analysis, *Journal of Building Engineering*, 44, 102704. https://doi.org/10.1016/j.jobe.2021 .102704

Patel, K., and Ruparathna, R. (2023) Life Cycle Sustainability Assessment of Road Infrastructure: A Building Information Modeling (BIM)-Based Approach,

International Journal of Construction Management, 23(11), 1837–1846. https://doi .org/10.1080/15623599.2021.2017113

Patyal, V.S., Sarma, P.R.S., Modgil, S., Nag, T., and Dennehy, D. (2022) Mapping the Links between Industry 4.0, Circular Economy and Sustainability: A Systematic Literature Review, *Journal of Enterprise Information Management*, 35(1), 1–35. https://doi.org/10.1108/JEIM-05-2021-0197

Roy, S., Rautela, R., and Kumar, S. (2023) Towards a Sustainable Future: Nexus between the Sustainable Development Goals and Waste Management in the Built Environment, *Journal of Cleaner Production*, 415, 137865. https://doi.org/10.1016/ j.jclepro.2023.137865

Sachs, J.D., Lafortune, G., Fuller, G., and Drumm, E. (2023) *Implementing the SDG Stimulus: Sustainable Development Report 2023*, SDSN and Dublin University Press. https://doi.org/10.25546/102924

Scrucca, F., Ingrao, C., Barberio, G., Matarazzo, A., and Lagioia, G. (2023) On the Role of Sustainable Buildings in Achieving the 2030 UN Sustainable Development Goals, *Environmental Impact Assessment Review*, 100, 107069. https://doi.org/10 .1016/j.eiar.2023.107069

Sharma, G.D., and Kharbanda, A. (2023) *The Role of Green Infrastructure in Biodiversity Conservation: Task Force 6 Accelerating the SDGs – Exploring New Pathways to the 2030 Agenda*, Observer Research Foundation. https://www.orfonline .org/research/the-role-of-green-infrastructure-in-biodiversity-conservation/

Sharma, R., Jang, J.G., and Hu, J.W. (2022) Phase-Change Materials in Concrete: Opportunities and Challenges for Sustainable Construction and Building Materials, *Materials*, 15, 335. https://doi.org/10.3390/ma15010335

Tichenor, M., Merry, S.E., Grek, S., and Bandola-Gill, J. (2022) Global Public Policy in a Quantified World: Sustainable Development Goals as Epistemic Infrastructures, *Policy and Society*, 41(4), 431–444. https://doi.org/10.1093/polsoc/puac015

Trane, M., Marelli, L., Siragusa, A., Pollo, R., and Lombardi, P. (2023) Progress by Research to Achieve the Sustainable Development Goals in the EU: A Systematic Literature Review, *Sustainability*, 15, 7055. https://doi.org/10.3390/su15097055

UN (United Nations) (2019a) *United Nations Environment Assembly of the United Nations Environment Programme, Fourth Session, Nairobi, 11–15 March 2019: Innovative Pathways to Achieve Sustainable Consumption and Production UNEP/ EA.4/L.2*, UN. https:// papersmart .unon .org/ resolution/ uploads/ k1900824 .pdf #overlay-context=node/243

UN (United Nations) (2019b) *Report of the Secretary General on SDG Progress 2019 Special Edition*, UN. https:// su stainabled evelopment .un .org/ content/ documents/ 24978Report_of_the_SG_on_SDG_Progress_2019.pdf

UN (United Nations) (2021) *The Sustainable Development Goals Report 2021*, UN. https://unstats.un.org/sdgs/report/2021/

UN (United Nations) (2022) *The Sustainable Development Goals Report 2022*, UN. https://unstats.un.org/sdgs/report/2022/

UN (United Nations) (2023a) *The Sustainable Development Goals Report 2023*, UN. https://dashboards.sdgindex.org/chapters

UN (United Nations) (2023b) *Seventieth Session Agenda Items 15 and 116 Resolution Adopted by the General Assembly on 25 September 2015, 70/1 Transforming Our World: The 2030 Agenda for Sustainable Development*, UN. https://www.un.org/ga/ search/view_doc.asp?symbol=A/RES/70/1&Lang=E

UNEP (United Nations Environment Programme) (2023) *Global Sustainable Development Report: Advance, Unedited Version.* https://sdgs.un.org/sites/default/files/2023-06/Advance%20unedited%20GSDR%2014June2023.pdf

UNFCCC (United Nations Framework Convention on Climate Change) (2023) *FCCC/CP/2011/9/Add*, UN FCCC. https://unfccc.int/files/adaptation/cancun_adaptation_framework/national_adaptation_plans/application/pdf/decision_5_cp_17.pdf

Wen, B., Musa, S.N., Onn, C.C., Ramesh, S., Liang, L., Wang, W., and Ma, K. (2020) The Role and Contribution of Green Buildings on Sustainable Development Goals, *Building and Environment*, 1851, 07091. https://doi.org/10.1016/j.buildenv.2020.107091

Westerholm, N. (2020) *State of Play for Circular Built Environment in Europe*: *Sustainable Building and Construction Programme Regional Report for Europe – Final Report June 2020*, United Nations One Planet Network Sustainable Building and Construction Programme. https://www.oneplanetnetwork.org/sites/default/files/from-crm/europe_final_201030.pdf

Wieser, A.A., Scherz, M., Maier, S., Passer, A., and Kreiner, H. (2019) Implementation of Sustainable Development Goals in Construction Industry: A Systemic Consideration of Synergies and Trade-Offs, *IOP Conference Series: Earth and Environmental Science*, 323, 012177. https://doi.org/10.1088/1755-1315/323/1/012177

Wiik, M.K., Homaei, S., Henke, L., Fufa, S.M., and Knoth, K. (2023) A Comparative Assessment of Building Sustainability Schemes in Norway, *IOP Conference Series: Earth and Environmental Science*, 1196, 012045. https://doi.org/10.1088/1755-1315/1196/1/012045

APPENDIX: INDICATORS (VERBATIM) FOR EACH OF THE SDGS

Table 5A.1   *Primary and secondary indicators for circular built environment mapped against the Sustainable Development Goals*

European Union	
Primary indicator	Secondary indicator
8.4.1 Material footprint, material footprint per capita, and material footprint per GDP	13.2.1 Number of countries with nationally determined contributions, long-term strategies, national adaptation plans and adaptation communications, as reported to the secretariat of the United Nations Framework Convention on Climate Change
12.2.1 Material footprint, material footprint per capita, and material footprint per GDP	11.6.1 Proportion of municipal solid waste collected and managed in controlled facilities out of total municipal waste generated, by cities
12.5.1 National recycling rate, tons of material recycled	7.2.1 Renewable energy share in the total final energy consumption
11.c.1 Proportion of financial support to the least developed countries that is allocated to the construction and retrofitting of sustainable, resilient and resource-efficient buildings utilizing local materials; a suitable replacement indicator is under development to be proposed for the 2025 comprehensive review*	11.1.1 Proportion of urban population living in slums, informal settlements or inadequate housing
9.4.1 CO_2 emission per unit of value added	12.a.1 Installed renewable energy-generating capacity in developing countries (in watts per capita); updated from Amount of support to developing countries on research and development for sustainable consumption and production and environmentally sound technologies in the earlier version*

Australia

Primary indicator	Secondary indicator
8.4.1 Material footprint, material footprint per capita, and material footprint per GDP	11.c.1 Proportion of financial support to the least developed countries that is allocated to the construction and retrofitting of sustainable, resilient and resource-efficient buildings utilizing local materials; a suitable replacement indicator is under development to be proposed for the 2025 comprehensive review* 6.4.1 Change in water-use efficiency over time
12.5.1 National recycling rate, tons of material recycled	7.1.2 Proportion of population with primary reliance on clean fuels and technology
12.2.1 Material footprint, material footprint per capita, and material footprint per GDP	12.7.1 Number of countries implementing sustainable public procurement policies and action plans
11.6.1 Proportion of municipal solid waste collected and managed in controlled facilities out of total municipal waste generated, by cities	
	9.4.1 CO_2 emission per unit of value added
	11.7.1 Average share of the built-up area of cities that is open space for public use for all, by sex, age and persons with disabilities 7.3.1 Energy intensity measured in terms of primary energy and GDP 11.1.1 Proportion of urban population living in slums, informal settlements or inadequate housing 14.1.1 (a) Index of coastal eutrophication; and (b) plastic debris density*

Africa

Primary indicator	Secondary indicator
12.5.1 National recycling rate, tons of material recycled	8.4.1 Material footprint, material footprint per capita, and material footprint per GDP
11.c.1 Proportion of financial support to the least developed countries that is allocated to the construction and retrofitting of sustainable, resilient and resource-efficient buildings utilizing local materials; a suitable replacement indicator is under development to be proposed for the 2025 comprehensive review*	9.4.1 CO_2 emission per unit of value added

6.3.1 Proportion of domestic and industrial wastewater flows safely treated	11.6.1 Proportion of municipal solid waste collected and managed in controlled facilities out of total municipal waste generated, by cities
6.4.1 Change in water-use efficiency over time	11.1.1 Proportion of urban population living in slums, informal settlements or inadequate housing
13.2.1 Number of countries with nationally determined contributions, long-term strategies, national adaptation plans and adaptation communications, as reported to the secretariat of the United Nations Framework Convention on Climate Change	7.2.1 Renewable energy share in the total final energy consumption
12.2.1 Material footprint, material footprint per capita, and material footprint per GDP	7.b.1 Installed renewable energy-generating capacity in developing countries (in watts per capita)*
	13.3.1 Extent to which (i) global citizenship education and (ii) education for sustainable development are mainstreamed in (a) national education policies; (b) curricula; (c) teacher education; and (d) student assessment
	11.7.1 Average share of the built-up area of cities that is open space for public use for all, by sex, age and persons with disabilities
	11.a.1 Number of countries that have national urban policies or regional development plans that (a) respond to population dynamics; (b) ensure balanced territorial development; and (c) increase local fiscal space*
	12.a.1 Installed renewable energy-generating capacity in developing countries (in watts per capita); updated from Amount of support to developing countries on research and development for sustainable consumption and production and environmentally sound technologies in the earlier version*
	7.3.1 Energy efficiency: energy intensity in terms of primary energy and GDP
	11.b.2 Proportion of local governments that adopt and implement local disaster risk reduction strategies in line with national disaster risk reduction strategies

	7.1.2 Proportion of population with primary reliance on clean fuels and technology
	7.a.1 International financial flows to developing countries in support of clean energy research and development and renewable energy production, including in hybrid systems*
	12.8.1 Extent to which (i) global citizenship education and (ii) education for sustainable development are mainstreamed in (a) national education policies; (b) curricula; (c) teacher education; and (d) student assessment 12.7.1 Number of countries implementing sustainable public procurement policies and action plans
	1.4.1 Proportion of population living in households with access to basic services
	2.4.1 Proportion of agricultural area under productive and sustainable agriculture

Asia

Primary indicator	Secondary indicator
11.c.1 Proportion of financial support to the least developed countries that is allocated to the construction and retrofitting of sustainable, resilient and resource-efficient buildings utilizing local materials; a suitable replacement indicator is under development to be proposed for the 2025 comprehensive review*	6.4.1 Change in water-use efficiency over time
8.4.1 Material footprint, material footprint per capita, and material footprint per GDP	11.1.1 Proportion of urban population living in slums, informal settlements or inadequate housing
12.2.1 Material footprint, material footprint per capita, and material footprint per GDP	12.7.1 Number of countries implementing sustainable public procurement policies and action plans
12.5.1 National recycling rate, tons of material recycled	13.2.1 Number of countries with nationally determined contributions, long-term strategies, national adaptation plans and adaptation communications, as reported to the secretariat of the United Nations Framework Convention on Climate Change

	11.b.1 Number of countries that adopt and implement national disaster risk reduction strategies in line with the Sendai Framework for Disaster Risk Reduction 2015–2030
	11.7.1 Average share of the built-up area of cities that is open space for public use for all, by sex, age and persons with disabilities
	12.1.1 Number of countries developing, adopting or implementing policy instruments aimed at supporting the shift to sustainable consumption and production
	17.18.1 Statistical capacity indicator for Sustainable Development Goal monitoring
	6.3.1 Proportion of domestic and industrial wastewater flows safely treated
	7.1.2 Proportion of population with primary reliance on clean fuels and technology
	11.6.1 Proportion of municipal solid waste collected and managed in controlled facilities out of total municipal waste generated, by cities
	12.a.1 Installed renewable energy-generating capacity in developing countries (in watts per capita); updated from Amount of support to developing countries on research and development for sustainable consumption and production and environmentally sound technologies in the earlier version*

Latin American and the Caribbean

Primary indicator	Secondary indicator
12.5.1 National recycling rate, tons of material recycled	12.a.1 Installed renewable energy-generating capacity in developing countries (in watts per capita); updated from Amount of support to developing countries on research and development for sustainable consumption and production and environmentally sound technologies in the earlier version*
12.2.1 Material footprint, material footprint per capita, and material footprint per GDP	6.3.1 Proportion of domestic and industrial wastewater flows safely treated
8.4.1 Material footprint, material footprint per capita, and material footprint per GDP	11.1.1 Proportion of urban population living in slums, informal settlements or inadequate housing

9.4.1 CO_2 emission per unit of value added	12.8.1 Extent to which (i) global citizenship education and (ii) education for sustainable development are mainstreamed in (a) national education policies; (b) curricula; (c) teacher education; and (d) student assessment
11.c.1 Proportion of financial support to the least developed countries that is allocated to the construction and retrofitting of sustainable, resilient and resource-efficient buildings utilizing local materials; a suitable replacement indicator is under development to be proposed for the 2025 comprehensive review*	7.1.2 Proportion of population with primary reliance on clean fuels and technology
	7.2.1 Renewable energy share in the total final energy consumption
	11.b.1 Number of countries that adopt and implement national disaster risk reduction strategies in line with the Sendai Framework for Disaster Risk Reduction 2015–2030
	12.7.1 Number of countries implementing sustainable public procurement policies and action plans
	6.4.1 Change in water-use efficiency over time
	13.2.1 Number of countries with nationally determined contributions, long-term strategies, national adaptation plans and adaptation communications, as reported to the secretariat of the United Nations Framework Convention on Climate Change
	13.3.1 Extent to which (i) global citizenship education and (ii) education for sustainable development are mainstreamed in (a) national education policies; (b) curricula; (c) teacher education; and (d) student assessment
	9.b.1 Domestic technology development: proportion of medium and high-tech industry value added
	11.6.1 Proportion of municipal solid waste collected and managed in controlled facilities out of total municipal waste generated, by cities

	11.b.2 Proportion of local governments that adopt and implement local disaster risk reduction strategies in line with national disaster risk reduction strategies
	7.a.1 International financial flows to developing countries in support of clean energy research and development and renewable energy production, including in hybrid systems*
	1.4.1 Proportion of population living in households with access to basic services
	5.5.2 Proportion of women in managerial positions
	8.3.1 Proportion of informal employment in total employment, by sector and sex
	9.5.1 Technological capabilities: research and development expenditure as a proportion of GDP
	11.3.1 Ratio of land consumption rate to population growth rate*
	15.1.2 Proportion of important sites for terrestrial and freshwater biodiversity that are covered by protected areas, by ecosystem type
	15.a.1(a) Official development assistance on conservation and sustainable use of biodiversity; and (b) revenue generated and finance mobilized from biodiversity-relevant economic instruments

Note: * Denotes a change from original wording of targets and/indicators subsequent to the research undertaken.

6. SBC achievements and challenges

Usha Iyer-Raniga

INTRODUCTION

The SBC programme focused on circular built environments as an area of interest a few years into the programme cycle. This focus was deliberate; it was designed as the programme found a gap in the market and saw the importance of a lack of engagement in the opportunities arising from transitioning to circular built environments.

The previous chapter presented the method used and the resulting findings arising from the mapping undertaken on circular built environments and the SDGs. This chapter presents the results of the primary research on stakeholders' views of the SBC programme. The programme lasted seven years, so the timeframe that stakeholders participated in was reasonable. There were no programmatic funds to run the programme, as explained in Chapter 4. In-kind resources were mainly in the form of personnel time and commitment to develop varied collateral (state-of-play reports and deep dives into circular built environments) arising during the seven years of the programme, apart from the 10YFP Trust Fund received for the programme within the first couple of years (also described in Chapter 4).

THE PROCESS

The authors wanted to get a balanced view of the range of different stakeholder perspectives on the SBC programme. The programme had 101 network partners, 24 MAC members, three organisations across the leadership team, five personnel from the Secretariat and the support of three board members and 200+ (OPN) National Focal Points by the end of the first period in 2022.

It was unrealistic to get views from all stakeholders connected to the programme. Also, it was not possible for all stakeholders to participate in the interviews, considering the in-kind commitment and the various time zones. The 10YFP Board and MAC members were from different parts of the world. Therefore, a selection was made to ensure the various stakeholder groups were represented in the feedback. The full set of stakeholders, their role(s) in the

SBC programme and details of those contacted are provided in Tables 6.1 and 6.2.

Table 6.1   Stakeholders interviewed to seek feedback

Stakeholders	Total	Interviewed
Leads	7	3
MACs	24	13
Partners	101	6
Secretariat	10	2
National Focal Points (NFPs)	200+ across OPN	1
Other OPN programmes (SPP)	5	1
Total		26

Table 6.2   Coding of the interview participants

No	Type of organisation	Code	Stakeholder
1	Government	GOV1	Lead
2	Government	GOV2	Lead
3	Government	GOV3	NFP
4	Government	GOV4	Partner
5	Government	GOV5	Partner
6	Government	GOV6	MAC
7	Government	GOV7	MAC
8	Government	GOV8	SPP
9	IGO (Inter-governmental organisation)	IGO1	Lead
10	IGO	IGO2	Secretariat
11	IGO	IGO3	Secretariat
12	IGO	IGO4	MAC
13	IGO	IGO5	MAC
14	Academia	UNI1	Partner
15	Academia	UNI2	MAC
16	Industry	IND1	Partner
17	Industry	IND2	MAC
18	Industry	IND3	MAC
19	Industry	IND4	MAC

No	Type of organisation	Code	Stakeholder
20	Industry	IND5	MAC
21	CSO (Civil society organisation)	CSO1	MAC
22	CSO	CSO2	MAC
23	CSO	CSO3	MAC
24	CSO	CSO4	Partner
25	CSO	CSO5	MAC
26	CSO	CSO6	Partner

Once the stakeholder groups were identified, questions were drafted to ensure their feedback on the programme was sought. There were some questions that were common across the programmes and some that were drafted particularly to obtain specific types of feedback relevant to that stakeholder group. Questions were high-level but, in some cases, the responses were drilled further depending on the stakeholder group and the role of the participant or their relationship to the SBC programme. Ethics approval was sought to ensure that the participation was voluntary and the engagement and interviews with the stakeholders did not breach any ethical procedures. The full sets of questions asked across the various stakeholder groups are presented in the Appendix.

Interview Respondents

The interviews usually took an hour and, in some instances, went on for over an hour. The interview questions (see the Appendix for details) were provided to participants prior to their signing their consent forms. At the start of every interview, consent was verbally sought again from the participants to affirm their interest in providing feedback. The interviews were recorded and transcribed. As the feedback on the SBC programme was to canvas views on the successes and challenges, nature of the engagement and governance arrangements, the questions were high-level but, where appropriate, participants were asked to focus on details.

Subsequent analysis was undertaken from a thematic perspective to understand views. The total of 26 participants from 23 organisations provided a multi-stakeholder understanding of their expectations of being involved in the SBC programme from 2015–2022, and whether there was value added from their perspective of being engaged with the programme. The number of participants interviewed provided saturation regarding their feedback on the SBC programme.

FINDINGS

The interviews analysis commences with high-level findings before focusing on the details. For ease of communication, they have been divided into key factors that have impacted the SBC programme. The analysis was undertaken thematically.

SBC Areas of Focus

The SBC programme focused on circularity from 2017. Embodied carbon may be considered to be the 'other side of the coin of circularity' in buildings and construction. Naturally, as more resources are used for buildings and construction, including infrastructure in the Global South, the focus on embodied carbon increases, as it plays a key role in construction. Embodied carbon is linked to construction materials, as buildings may be designed to also incorporate carbon sinks. This is particularly important in the Global South as many buildings have yet to be built and the interest of the Global South is in rapid delivery of the built environment to meet urgent needs, whereas the Global North has had the luxury of being able to reflect on the impact of the built environment; hence circularity has received attention. For too long, there has been a focus only on the operational energy of buildings, and as this is getting more stringent, it is essential to also consider embodied energy. The same high-impact building materials are being used, and therefore new materials on the market are essential.

Participants highlighted a number of important considerations regarding sustainable buildings and construction as a matter of urgency. As buildings are becoming more efficient, embodied carbon is an issue. Respondents highlighted the importance of:

- embodied carbon and the full life cycle of buildings
- the value chain and the impacts across it incorporating building waste and emissions associated with it – emissions to land, air and water
- the role and importance of circular economy
- a specific focus on materials, as materials determine the embodied carbon and associated emissions across the life cycle
- production systems of materiality and construction through to management of the end of life of buildings
- SMEs and local considerations, including local materials, jobs and skills. The more products that are imported, the higher the embodied energy, which is not where the industry wants to land.

Leadership opportunities were raised, such as:

- collective action to ensure embodied carbon is part of the narrative going forward
- developing a common language
- communication with the industry and various stakeholders, including clients
- breaking silos to remove current fragmentation
- establishing a yardstick to measure, and agreement on the measurements: building codes, standards, certifications and tools to support circularity.

This is highlighted by a quote from one of the MACs, CSO5:

> Embodied carbon and circularity represent the whole life cycle. This is important not only in terms of resource efficiency and of improved tracking of carbon, but also in terms of changing the whole value chain mentality because it forces value chain participants to actually move closer together, communicate much better and share data. As I see it, the resource aspect is definitely crucial, but in addition there is also the real game changer of breaking down silos and making stakeholders not only think about their life cycle stage and their own interests but also that of others. So for me it is a win–win on both sides if successfully implemented.

And from the Secretariat, IGO2:

> The easiest way to avoid pollution is reduce material throughput and avoid polluting or toxic inputs in the first place. These are no-brainers which policymakers very regularly miss.

CE focus

Circularity sets the tone for the use of construction materials and the impact of materials over life cycles. In developing countries (but also in many developed countries) governments are faced with the issue of construction, construction materials and building waste. New jobs and skills can be opened up, taking the circularity perspective. The climate argument needs to be emphasised in this context. Circularity is holistically about carbon, not about waste. Circularity needs to be linked to the broader agenda.

Understanding of what materials are made of, their provenance, reuse and other R-strategies (refuse, rethink, reduce, reuse, repair, refurbish, remanufacture, repurpose, recycle and recover [waste to energy]; refuse being the best strategy compared to recover) at end of life, procurement processes and the impact on the construction sector and its end-of-life options, covering scope 1, 2 and 3 emissions, is needed. A huge demand gap in circularity exists between the Global North and Global South. Materials use, the mode(s) of manufacturing, policy context and praxis, and the value of materials use, all need to be considered.

There has been a very long trajectory of energy efficiency, and operational energy perspectives of buildings in use are just coming into discussion in the Global South as the embodied energy considerations are being recognised due to the sheer volume of construction taking place. It is just not feasible to continue to use linear models.

The responses regarding the importance of circularity in the built environment were highlighted through these emergent themes:

- the urgency of the issue and its critical role in the future
- the need to have clear definitions, and also interpretation by people, data, indicators, metrics, tools, policy measures and legislation/regulation
- context-specific interpretation
- the importance of local, and the role of SME and grass-roots engagement
- natural resource use
- new business models
- new ways of living
- producing products and materials with low emissions
- incorporating and understanding the role of supply chains
- transportation of materials
- biodiversity and water and other resource implications at local and regional levels
- chemicals used in the material context, especially if materials have to be reused more than once.

Circularity is applied globally; it is a model that can be applied to the rest of the world. But the use of local materials is critical as they are available within close range, and if natural materials are used, they become part of the biological cycle again. Early vernacular systems are almost non-existent now and need to be revived. Designs, forms and structures need to be examined. Low-rise construction may still work in the modern world of telecommuting. The Global South is heading to taller and taller buildings that may need to be rethought to smaller scales. Respondents noted this in their feedback:

> I think for me, it's an overall philosophy in a sense. (IGO1)

> We need to have circularity models in buildings, construction and other manufacturing processes. You can't just use a linear model for growth, so circularity has a huge role to play. And [in] most of the cases it needs to be reused, remanufactured, and the best thing is to refuse natural resources and reuse whatever is already extracted. (CSO2)

Upscaling CE and thinking in the SBC programme

Circularity should not just be for the top end of the market nor continue with the current focus on the developed world or regionally across the EU. The focus should be on bringing the whole world to a level playing field, using local building materials, and really understanding stocks and flows so that supply and demand may be balanced. The role of developed and developing countries in this context is important. In particular, the role of the private sector in developed countries is paramount, and they may be used as a template in developing countries, so that lessons learned may be applied in the developing-country context with no greenwashing. Low-tech options may be used. It is essential to lift off existing knowledge, rather than continuously reinventing the wheel. The role of stakeholders, regulation and building codes, the enforcement of regulations, capacity building and procurement all need to be brought into the mix through a holistic approach.

It is contradictory to see high-rise buildings, especially in developing countries, that have the best green building ratings when right next door to these high-performance buildings are slums. The mass market needs to be considered, not just the high end of the market. There are a full complement of issues, particularly around materials and supply chains: chemicals in materials that hinder reuse and repurposing, buildings as material banks, modular construction and other such approaches to support CE, capacity building, and the use of locally appropriate materials and local skills and technology.

There should be a focus on independent scientific research to build demonstration projects so that people can understand how to use local materials. The idea of local needs to be promoted where possible, and the use of local tools needs to be especially encouraged. Some international tools may support climate conditions which may be similar across countries, but they may not be able to replicate building materials or local construction technologies. Considering local people in the process of building and construction will ensure local jobs; this builds a sense of community, and then the private sector and the decision makers may be brought into the discussions to upscale the outputs and outcomes. It is essential to close loops at the micro level so that waste – for example, waste heat – can be incorporated into energy production.

The following points were highlighted:

- use of demonstration projects to replicate and scale up, and setting up guidelines so it is easy for others to follow
- use of knowledge to engage governments and seek their support, to show that circularity is of little or no cost
- the role of legislation for CBE – the policy directions are also critical as the context, construction practices and procurement processes are all different; both top-down and bottom-up approaches are needed

- the role of the labour market, and especially the SMEs, who need assistance the most – they need interventions where they can recoup costs in a short period of time as they cannot afford a lot of investment; modelling across business needs and returns on investments should be encouraged and used for decision making.
- the role of the end user; the customer and customer needs; affordability, quality, benefiting from environmental deterioration (e.g., sand mining)
- the use of data, metrics and tools is critical; circularity needs to be measured to enable monitoring and evaluation
- awareness and education, capacity building
- taxes and financial incentives based on the footprint and handprint created
- new approaches such as business models and business opportunities should be supported.

On replication and scaling up, IGO2 noted:

> There's a little phrase: replication and scaling up of sustainable policies and practices – this is central to the work of SBC.

IGO3 stated:

> When you rely on your local resources and local material for construction you can ensure more circularity because you are working in an ecosystem where materials can go back somehow. If you bring materials from outside that do not fit very well with the ecosystem you create more problems and pollution, and disruption.

Supporting a future in buildings and construction

A systemic change is needed that considers a holistic approach involving regulations, providing a level playing field, tax incentives and other such financial levers to finally create a system with diverse approaches. Once the system is created, it is possible for stringency to be increased over time. However, the big question remains as to how to apply circularity in practice, and innovate and stimulate markets, resulting in holistic approaches that bring in whole-life-cycle thinking, life cycle costing and the wider socio-economic aspects of resource use.

Collaboration, cooperation, sharing data and transparency of material composition and across the supply chain, reuse of materials, embodied content of material use, understanding the carbon footprint, business cases using data, procurement at local government level, education, awareness, and the infrastructure to support reuse in the market and the creation of new exchange markets for reselling goods and materials, all need to become mainstream, quickly.

To support a future in buildings and construction, respondents suggested the following:

- clarity in communication on circularity, as there is a lot of fragmentation now and it is quite ad hoc
- highlighting CE's importance and benefits and raising awareness of the issue, so that professionals, government officials and the general public understand what it really means
- clarifying technical standards, language and metrics, so everyone is 'singing from the same hymn sheet'
- education, so that people understand the capacities needed to support circular outcomes
- paying attention to the informal sector, as much as the formal sector, which needs different levers
- considering the materials perspective, especially in the Global South, as it has unique challenges in the form of new construction, bio-based materials and local materials, ensuring that materials are not imported (as this involves considerable scope 3 emissions and does not help with the embodied debate), and promoting local jobs for new construction in the Global South
- standardised tools – the building and construction industry is struggling with these concepts
- giving attention to the role of finance, as the industry is still struggling with this aspect – when circularity is added to it, it creates a spanner in the works
- realising that the role of government engagement is critical, along with local engagement through participation of local communities in green buildings and informal settlements.

With respect to the future of buildings and construction, MAC member IGO5 stated:

> In my perspective, the sustainability of the buildings and construction sector demands a thorough re-examination of regulatory policies, which currently lack a strong commitment to circular economy principles. While some sustainable technologies and passive design features are recognized, a deeper inspection reveals a need for more comprehensive policies that recognize the negative and profound impact of certain industries and their ongoing conventional practices in building operations. To address this, we must implement top-down approaches that prioritize circularity, such as expediting through regulations large companies' transition to more eco-friendly production methods, especially in industries like cement and steel manufacturing. Likewise, we also need to prioritise investments in small and medium-sized enterprises pioneering sustainable materials. This multifaceted strategy is crucial for steering the construction sector towards genuine sustainability.

This was also expressed by another MAC member, IND5:

> Let's put it that way, and the central part is the economic case, but if you are a government, one of the things you are keen on is more entrepreneurship. You are keen on more jobs if you are a developer. You know you are keen on a long-term revenue stream and maintenance, and entrepreneurs that assist in achieving that.

Promoting CE

Circularity can contribute to decolonisation, create better buildings via holistic thinking, set up clear business cases and build capacities through education and training. Members of some peak industry bodies are seeking a level playing field; there are barriers across the value chain that need to be closely monitored, and structural blockages should be addressed, such as management fees linked to overall costs of project and management of assets linked to budget allocations. Risks across the supply chains also need to be considered, especially where standards and certifications are not in place and enforced. Knowledge and experience are not commonplace, and markets have to be created to respond to certain needs for circularity, such as new infrastructure to create a secondary materials market.

The balancing of 'carrots and sticks' is needed to move the industry to consider CE practices. Regulatory interventions provide incentives so that the industry is able to make the required changes to its practices. In the example of one country provided by a partner, a national committee was formed that assisted in putting circularity on the map, then engaging with the right players in the country to catalyse and shift the transition. There needs to be private engagement through the 'big end of town' being involved and showcasing examples for leading circular thinking. Grants and funding are needed, but the lack of knowledge needs to be addressed:

- Metrics and tools: guidance on materials and other items, calculators, building passports, data transparency, understanding the building stock, simple tools for putting circularity into practice
- Education: educating the client, educating the industry, changing valuation processes to value existing buildings (beyond heritage value)
- Transparency and disclosure: so that there is an ability to compare 'apples with apples' supported by legislative measures to force disclosure
- Economic value-add: jobs, entrepreneurship, new business models, manufacturing and innovating new construction materials and techniques of construction
- Design: demonstrating value in the design now and in the future, considering the implications associated with waste implications, canvassing deconstruction opportunities

- Communication: getting the messaging right for all stakeholders, including the right messaging to government stakeholders
- Procurement processes: laying down new guidelines or rules for procurement so that procurement is not locked into linear cycles – instead the focus is on performance
- Holistic approaches: overcoming current fragmentation in the market and supporting shared goals and aligning across the various stakeholders in the value chain
- Learning from energy efficiency, which has had a long lead time, and moving beyond building codes to consider circular applications
- Practical examples: demonstrations and pilots are needed to showcase to all stakeholders, which requires funding and a willingness to trial and take the associated risks

The importance of promoting circular built environments has been expressed by a partner, CSO6:

> The local government has another idea about circularity or the embodied carbon idea and we need to be sure that we have a comprehensive framework to begin change, and another thing in this work is not only a common vision but realistic goals and change in strategies with all actors. When we talk about circularity it is about what we can do about it in three years and seven years or in ten years. We only talk about the issues but not the goals and local strategies to begin the change. Another thing is development and institutionalized instruments and tools to make transition possible and monitoring the trends … one of the main gaps that we found is the education gap because, for example, we still find that people leading the circular economy in the ministries, they think that circularity means recycling.

Another partner, IND1, highlighted the importance of transparency:

> Like for like comparisons are needed. And, you need measures and compatible protocols, but that is the other part … for example, in the construction material industry that is advancing with roadmaps to reduce emissions, there are no worldwide agreements or protocols yet to define and measure low-carbon materials. I believe that it can't be a decision of each country or each producer if we want to achieve real results. Today, you can find some companies offering low-carbon-emission materials but, when you ask yourself "in comparison with what", sometimes you will find that is in comparison with high averages of other countries or using a different or self-made protocol. This is happening because there is no clear regulation, everybody knows that some countries have specific bigger footprints for some products or there are some protocols that are more favourable. So that's an issue there from my point of view.

Yet another partner, CSO4, focused on the entrenched linear systems, stating:

> Change in procurement from units of materials to overall performance will have
> an impact on the design and performance of buildings. The standard multiple layer
> construction value chain doesn't incentivize performance. Performance can really
> be incentivized with the current construction project models between the client and
> the contractor. But in most of the cases, at least in my humble opinion ... we then
> lose that incentivization for performance when the contractor starts chopping down
> the job to various subcontracts and hundreds of materials and systems supply con-
> tracts, and these are all based on, you know, companies like mine actually earning
> on the basis of the amount of material we supply. And ... I know we could even
> compete better if the basis of competition wouldn't be the amount of material times
> unit cost. But if it would be, for example, performance-level embodied carbon and
> a couple of other things. So, we should find ways with which, for example, system
> providers like ourselves would not be compensated based on the amount of material
> we supply, but based on the performance of the system that we supplied. We can do
> ... the distribution system with a kilometre of pipe or 100 meters of pipe with the
> same performance level. If somebody specifies a kilometre system, we can't change
> that anymore. We are forced to supply a kilometre.

SBC Achievements

SBC has influenced awareness-building and created networks especially for
knowledge building and training, improving information and supporting educational policies. It has worked across government, industry, the private sector
and academia to drive change. Engagements at national and state government
levels have produced modest results. Financing and procurement changes
also need to go hand in hand to produce a sea change in the sector. The SBC
programme (MACs, CDs, network partners, the Secretariat) generally met at
least once a year. Targeted workshops and site visits that accompanied the
face-to-face meetings worked well. The SBC programme focused on one
major issue and did that really well: the focus on circular built environments.

A MAC member, GOV8, said of this achievement:

> I think this is something the SBC programme has done incredibly well compared to
> my experience on other programmes.

Mapping the SDGs

The SBC programme has responded to the SDGs and was the only OPN programme that explicitly worked on the SDGs, undertaking primary research.
All interviewees agreed that the OPN has definitely covered a gap in the
marketplace, especially in SBC. The literature demonstrates that there is not
much work done on circularity holistically or linking circularity to the SDGs,
especially in an empirical manner, as explained in Chapter 5. The surveys
undertaken across the SDGs were particularly important for this. The surveys

assisted in gathering primary data for understanding the link between SBC, the SDGs and CE in the built environment, highlighting the role and importance of circular built environments, and reinforced the programme's work and commitment through the surveys that were deployed. The SBC programme made a clear commitment and actively engaged with the SDGs.

All respondents agreed that there is no doubt that there was a value-add in undertaking the mapping to respond to the SDGs. The work on sustainability generally is linked to the SDGs, yet this has not necessarily been clearly articulated in the built environment sector, at neither a project nor building level. Ongoing monitoring helps in fine tuning and shaping the work associated with the SDGs. In some organisations that were part of the MAC, the engagement with the SDGs really helped them understand the practical challenges. A lot of the interest from some of these organisations was very EU-centric, so that was a limitation as they did not necessarily feel there was value in taking a global approach.

Some respondents stated that the SDGs are interlinked, and the starting point of which SDGs to commence with is, in a sense, not relevant. Many respondents noted that it is not just the SDG indicators but relevant data that is also needed as evidence. The SDGs may also be used for training. SDG 12 is central, but the engagement across other programmes of the OPN and the linkages they form has yet to be mined further. Across the SDGs, largely sectoral issues are considered, such as water, energy and food, resulting in SDG 12 – which has a combination of all of these – being largely overlooked. Systemic and holistic approaches are needed. SDGs are quite broad, but they are not perfect, and this is a general challenge with them.

Construction organisations are trying to respond to the SDGs and are not always clear on how this needs to be undertaken. The value chain of the buildings and construction sector should also respond to the SDGs. Demonstrating grounded results is critical. Showcasing through examples, demonstration projects and case studies, and experimentation, lead to transitions in knowledge and information and, ultimately, in achieving the SDGs.

Respondents reported that working on in-depth material circularity issues led by the Co-Leads of the SBC programme was a great experience. This approach provided a very good platform to raise issues, and understand where the bottlenecks are and how to think about presenting solutions. Understanding building and construction materials and waste was also useful. The importance of jobs to support the formal and informal economy was recognised by interviewees, particularly those working in the Global South. Those that worked in the Global North and South recognised that the use of recycled materials in the building code can be a problem as there needs to be quality assurance in the use of recycled materials. Operationalising circularity in the context of reuse of materials and quality assurance of structural integrity needs to be part of the

SDG and circularity narrative. If assurances of material reuse are not provided, these end up being imported from elsewhere, increasing the embodied energy across scope 1 and 2 emissions.

Circularity is difficult to achieve in practice in the Global South as there is suspicion that as the market is so unregulated and has a large informal sector, and that buildings that are made with reused materials will fall down and will be less safe. So, some tools to operationalise and conceptualise material reuse are helpful. Over time, this can become more rigorous and stringent.

Some interview participants that worked on practical applications indicated that the SDGs should have been mapped at a project level by undertaking a project and seeing how that responded to the SDGs. SDG 12 influences other SDGs, and to consider SDGs holistically a paradigm shift is needed. SDGs need to link to the human agenda; the people need to be placed at the centre. A MAC member undertook work on SDGs 12 and 11, particularly on targets 11.1 and 11.c on housing. Other respondents denoted the importance of the links between SDG 12 and others: SDGs 1, 2, 3, 5, 6, 8, 11, 13 and 17.

One of the Leads of the SBC programme, GOV1, underscored this achievement with this statement:

> Yes, I think that, for example, compared to other programs, … we have been addressing SDGs sufficiently in SBC, yes. And specially this study on circular built environment and connection to SDGs … this is something that nobody else has ever done. So, it's very good to see these results in SBC. But as I said, I think that we could have focused more on SDG 12 and the indicators further.

The importance of the SDGs was also echoed by the Secretariat, IGO3:

> SDGs 8, 9, 11, 12, 13 are critical for SBC. If we can cover mainly those five, starting with 11 and 12, adding the 8, 9 and the 13 you have covered 90 per cent of what we need in the sustainable buildings programme. SDG 12 has a very broad mandate, and this broadness begs for support. Most governments are yet not well aware of the challenges and opportunities. Clearly, these are lost opportunities.

Upscaling SBC

There was a lot of interest in and growing awareness of the SBC programme. Engagement with the network was established to make good use of the expertise already available, but scaling up was staggered.

Interviewees noted that to scale up, working with national governments is needed. A national policy that ties urban development with SBC is needed. Beyond national governments, it is critical to engage with the green building councils due to their reach. Respondents agreed that it is essential to move the dial from general agreements to specific projects. A holistic collaborative approach to change for circularity is needed, moving from a silo mentality to partnerships and engagement.

One of the interview respondents, IGO2, had this to say about the significance of upscaling the SBC programme:

> Because it's about construction of all the infrastructure that's going to be added to this fragile and degraded planet in the next 30 years. If that's unsustainable infrastructure, it's a total disaster for the planet. Total disaster!

The importance of upscaling and the focus on circularity in particular was reiterated by a MAC member, IGO4:

> Niche focus on circularity ahead of our time ... [The] SBC programme came on board much earlier than it should have, it differentiated the SBC programme from the other buildings and construction groups. But it's a little bit ahead of its time, I think. It doesn't get the government's attention as much. And, because it's not such a critical issue yet and it doesn't link as directly to climate finance as energy efficiency. And so ... I mean, of course it's going to become more and more important, so it's ahead of its time, I guess.

Networking

When asked if the SBC programme added value, all respondents expressed a positive sentiment in this regard. Some felt that it had settled into the circularity area at the right time, three years or so into the programme, as nobody was talking about CE then, especially in SBC and as part of the wider OPN discourse. There were precursors to CE of course, in the form of green economy, especially in the context of sustainable development. SBC had a clear and strong focus on circularity but it needed to move to the next step to determine the level(s) of influence and actors involved in the process to ensure there is no duplication.

Participants agreed that SBC stakeholders have definitely benefited from being part of the network. There was a definite value-add in engaging with the SBC programme as it was able to contribute not only to the global dialogue, but also at regional and in-country levels. It supported partnerships and collaboration that was otherwise not there, and opened gateways for collaboration across continents and countries. Face-to-face meetings assisted in engagements and building relationships. After COVID-19, this tapered somewhat as online meetings are not the best way to engage with each other to build relationships.

The SBC programme showed that engagement and networking can reap rewards if there is credibility associated with them. The engagement across the programme has helped some of the MACs to bring their expertise and 'get better bang for their buck'. Working across the OPN in some instances was also helpful, and engaging through networking worked with other OPN programmes even though there was limited engagement.

As a network it provided complementarity, and balanced various trends, especially between developed and developing countries. This was an eye-opener as the circularity focus for a long period of time was set largely in the EU context. It also brought in diverse interests across the programme: academics, government, international agencies, NGOs, international agencies, and the like. The sharing of knowledge and expertise was unparalleled. It was a two-way street where knowledge exchange was supported. Examples quoted were working on the state-of-play reports that assisted in developing knowledge and gaining expertise through engagements with respondents' organisations in other countries. The engagement allowed connections across various silos, which normally cause fragmentation, to make this a unique global programme.

Some of the interviewees pointed out that at some levels needs were different, yet at others they were similar, especially across the Global South. One point of difference for the SBC programme was the focus of the buildings and construction programme in the Global South, which many international platforms did not focus on, creating a clear value-add. It offered a richness, particularly beyond the Global North perspective. Network members have an opportunity to leapfrog into the future. Participants said that it was very encouraging and motivating to see some examples of practical applications coming to the table. Some respondents indicated that it is always the case of 'preaching to the converted', and the target audience needed to engage those that are typically not part of the circular conversation in the buildings and construction sector.

The overall network, projects, people, diversity of approaches to achieve circularity and the structures overall: the jigsaw of how various structures fit together and the levers that are needed to be pushed or pulled in different contexts contributed to increased understanding. Networking amongst the MACs in particular brought in the exchanging of ideas and information, even with online meetings. Face-to-face meetings were, of course, better than online.

All respondents agreed that the SBC programme has been very helpful in bringing diverse groups of people together. Messages were not skewed when this group came together and worked towards finding solutions. Experts were at hand when needed. The resulting state-of-play reports did not just become knowledge products that were valuable, but the inclusion of the case studies provided other useful products with grounded solutions for industry and government. Not only were there outputs in the form of various publications, but the ability to bring in peak agencies and work across borders and regions demonstrated that collaboration works well.

The added value of the SBC programme has been underlined by the following respondents representing the Secretariat and MACs respectively:

Absolutely, SBC has added value. (IGO2)

OPN and SBC is a landing space for many things but needs to be well supported. (CSO1)

I think the projects have been very useful, so the publications and the kind of collaborative working has been useful ... the sharing more broadly with other parties with some of the UN agencies, some of the non-governmental organizations, the donor agencies and then the governments and civil society in different countries. (IND5)

SBC as driver of circularity and embodied carbon

When asked how a global programme such as the SBC programme can assist in reaching circularity and embodied carbon goals, the responses from the interviewees were all positive. Again, all the respondents indicated that the SBC programme achieved modest outcomes without financial inputs. It has an international presence that can provide credibility to the work undertaken. The role of research and documentation (through the state-of-play reports and other publications) is essential to express the importance of a decarbonised economy and focus on circularity. Participants noted the importance of circularity beyond waste management percolating slowly through countries, particularly in the developing world. Building and construction waste is a problem that needs to be put onto the agenda, and the narrative needs to include waste management but also incorporate circularity principles beyond it.

The programme has already assisted countries, organisations and the sector in a number of different ways:

- It assisted in operationalising and implementing solutions, however limited their impact was at the time (such as Trust Fund projects and state-of-play reports).
- One MAC member highlighted that through their engagement in a state-of-play report, they were also able to work in various countries (Burkina Faso, Senegal, Rwanda and Uganda), raising the level of interest but also looking at ways of operationalising codification of circularity and green buildings.
- Another MAC member commented on the diverse community of practice, best practice examples, policy analysis and briefings, and the attention paid to developing the evidence base through research (case studies), highlighting cost savings and material efficiency benefits. All these have added to the value provided in what the SBC programme did.
- All participants stated that the SBC programme assisted in making connections, and the network supported discussions between people from

different kinds of backgrounds. Sharing case studies and lessons drawn to support other cases assisted knowledge sharing/education and raised awareness.

- As a network, the SBC programme benefited other projects, and it benefited from them, according to most of the respondents. It was a great 'two-way street'. The programme strengthened relationships in countries and with various UN agencies. It improved relations with governments, but government departments are not consistent over time, and that in turn has its challenges. Structural relationships need support to effect change.

Going forward, respondents suggested that:

- Making change to circularity happen practically is not easy, so the innovation and opportunities to scale up globally require some strategic thinking. For instance, how best to use international networks across the OPN, and how best to highlight examples of building back greener.
- SBC needs to focus more on communicating findings and use the positive feedback loops to make an overall positive impact. The SBC programme did not invest enough in marketing.

The role and benefits of the SBC programme have been highlighted by various respondents:

> But as a program or a network I think the most important thing is knowledge sharing and getting there together. The actors together sharing information create a source of knowledge, not only data, but the knowledge itself is shared. (GOV6)

> I think those global international platforms are very important. The science community really has a very important role to play because it's actually the science that needs to provide evidence on our challenges and solutions. (IGO1)

> OPN provides knowledge and information – I don't know how to keep those people or to keep that knowledge and to keep that networking. Being in [the] One Planet Network, for my point of view, you can find a lot of information. (IND1)

Missed opportunities

The issue of missed opportunities was not deliberately raised as a question with the interview participants. However, when the interviews took place, some of the participants who were part of the programme from the beginning reflected on issues that could have produced greater impact if a little more effort had been expended as part of the programme. These reflections involve slight approaches or modifications that would have made the programme more efficient, produced impactful results and scaled the impact. The intent of this subsection is broad, focusing on various aspects of the SBC programme, including governance and funding.

These are highlighted below:

- The SBC programme took a while to find its niche. This reflection came from a participant who felt that it found its 'groove' after a few years. This highlights the nature of the global engagement where almost all the participants were not known to each other at the start of the programme and needed to build relationships with each other as well as deliver on its key performance indicators (KPIs).
- Another participant commented that time needs to be spent to understand the resource mobilisation sector a bit more to hone in on where money should be spent. That is, mapping was needed to work out who was doing what, and which donors were working in the SBC space.
- The language/taxonomy and terms associated with circularity in the built environment need attention. A common understanding is absent, but it would have been possible to develop a language of circularity once it was explained properly.
- SBC programme stakeholders met at the MAC meetings but very few followed through later on to strengthen the network and engagements with each other.
- The survey undertaken for the SDGs did not initially resonate with the Global South as the industry experts did not know about circular built environments or the SDGs, and the experts needed more guidance in developing countries as they were unable to tie in the SDG indicators with the engagement in their own regions.
- With the supply chains, there was little focus on the private sector and little engagement across the demand side of circular built environments. Some participants noted that this is particularly relevant when dealing with national governments.
- Fragmentation across the OPN did not help. If there had been more effort made across other OPN programmes, it may have assisted in taking some collective effort to address SDG 12, which has had relatively few funds devoted to it. While there was limited engagement across the SPP, SLE and STP programmes, some participants felt that the Secretariat should have focused more on building bridges across the programmes. Even the funding that eventuated from the OPN MPTF took a siloed approach across the programmes. Each of the programmes worked in their own silos and it was difficult to engage across the programmes, particularly as some of them worked on common issues such as water, energy and other topics. If the programmes worked collectively, this may have also assisted in collaborations and helped to widen and deepen the impact of the SBC programme.

- While case studies were collected, presented and disseminated, there was no further work undertaken to mine them to develop them into practical tools and business cases or use them in targeted training. This would have been very helpful.
- While the importance of vernacular and indigenous approaches was highlighted in the programme's work, there was no specific guide/tool developed or publication on this, save one undertaken towards the end of the programme in Africa as part of the deep-dive state-of-play report. Participants said that those simple principles of good design that people have been using for many thousands of years would have been helpful, especially in the emerging markets.
- The UN agencies in countries, especially where work was undertaken by the SBC programme, could have acted as a binder and supported translation on the ground and also vertical integration across governance structures. This point was raised by participants who work closely with some of the UNCTs.
- There was no effort made to market the work undertaken. While it was no doubt significant, communication in simple language and the use of infographics and such forms of communication would have helped. This is despite the programme developing podcasts and simple publications to understand case studies and link these to the SDGs. Marketing should have been built into the programme both at the Trust Fund project stage and at the capacity-building point of tools such as SHERPA, sometimes beyond the SBC programme, with the delivering agencies across the Trust Fund projects (such as UN-Habitat or UNOPS). If marketed well, collateral would have been 'out there' to be used beyond the countries where the programme was trialled, broadening and deepening project outcomes.
- Most of the communication was in English, but from a global perspective the English-speaking world is only a proportion of the total.
- The programme's work in the emerging markets was useful, especially the case studies. However, this was not highlighted beyond its website and sporadic presentations at international conferences. They could have been mined for training.

Some of the MAC respondents, a Lead and one partner offered the following reflections:

> I believe reports … it's useful, but most of the time people just look to read the summary and that's all. (IND3)

> It couldn't have been perfect because unfortunately perfect doesn't exist … For me it's very much a question of indicators, monitoring and evaluation and having that structure a little bit more clearly established so that we are clear also on what we

can provide. The reporting system should not be too complex and too cumbersome, because I think the previous one was quite demanding but not very effective. (IGO1)

Need to translate into local language. The information always or many times is only in English or German and the many people in developing countries don't speak more than their native language. That was one point during the implementation. For example, for our certification system here, we offered a helpdesk in Spanish and that helped a lot. To move to the circular economy, you need to work with the people. (IND1)

SBC Challenges

Engagement challenges

Participants agreed that generally there was good engagement across the SBC programme. The MAC members met at least once a year and were able to bring in their knowledge and expertise and share meaningful discussions over projects and other outputs associated with it. Participants reported that meeting agendas were clear, so they could come to the meetings well prepared and knew what to expect at them. MAC members and other participants had an opportunity to get to know each other and connect with their peers post meeting, if desired.

Reports prepared by the programme were reliable and useful. The network has been used to harness various perspectives to fill in knowledge gaps and has set up a foundation of reliable information; the availability of go-to people to respond to gaps in information resulted in organisations engaging with the SBC programme to meet their own goals and derive value from engaging with some of the MAC organisations. It has assisted in moving beyond countries and crossing borders.

A focus on the building scale alone brought challenges. Natural scaling to precincts and to the built environment as a whole, including extending the work to focus on neighbourhoods and infrastructure, would have also been helpful.

There was no funding for the MAC members to engage with the programme, so commentary on collective engagement activities that were expected as part of the OPN, such as reporting, was an issue. The value-add of reporting on the OPN platform was unclear. In some instances, the MACs reported that the annual reporting cycles got their teams across different countries excited about reporting on the work undertaken during that reporting period.

There were limitations on what more could have been achieved from the MAC meetings beyond the face-to-face and online meetings. Maybe people in certain regions could have continued engagements at regional levels, but to do so needed commitment and, of course, time and resources for travel.

Face-to-face meetings as part of other events (WUF, World Circular Economy Forum [WCEF], COP, UNEA, etc.) assisted, as duplicate travel was avoided.

Focusing not just on the technical aspects but also cultural and other aspects of the SBC programme was also critical. Information was generally lacking, so more demonstration projects would have assisted. These projects would have been able to encompass parts of building design, construction and deconstruction, so it would have been valuable to show them to industry.

Not having a common vision across all the OPN programmes, especially on common indicators such as the SDGs, and not driving the programmes at the same speed did not help. Some respondents indicated that the global structure of the SBC programme was too high-level; they felt that it was difficult to engage with and they could not link the outputs and outcomes at the local or sub-national levels. It was essential to have a balance between focus and diversity of approaches at a project level. Almost all respondents felt there was insufficient engagement from the OPN Secretariat and CDs regarding functions such as cross-programme collaboration, resource mobilisation, and so on, despite some very strong relationships across the Network's SBC MACs.

Often, the person representing an organisation as a MAC member or partner committed their time to the SBC programme and the relationship was built with this person. When the person left the organisation, they were not replaced, for several reasons (organisations not committing to a resource, the new person not being interested, changing priorities, etc.), resulting in no ongoing relationships in some instances. The engagement was not always organisation-to-organisation even though it may have been so at the start of the SBC programme, demonstrating that the people-to-people connections and relationships were important as well.

The general consensus was that the GlobalABC has had more visibility and traction compared to the OPN SBC programme. As the network members across the two platforms overlapped, the programme ended up taking a back-stage position despite commencing before the GlobalABC came into the picture.

The focus of the SBC programme on circularity came in at some point along the way and members of supranational bodies recognised its role, importance and fiscal considerations. Bringing in international bodies such as the World Green Building Council (WorldGBC) to continue to engage with the SBC programme would have helped, as the WorldGBC is a unifying body for the green building councils and it has a direct reach into the countries where these councils are in existence, particularly in the Global North, where many country-based green building councils are in operation. There are a larger number of such councils in the Global North than in the Global South, and therefore the representation of the Global South in this international organisation is also limited.

The National Focal Points had very little to do with the OPN. It would have been useful to have OPN programmes meet with the focal points and discuss strategies on a regular basis. At the national level, the focal points represented government interests, and this would have nurtured government interventions if efforts had been invested in these relationships. Likewise, the 10YFP Board also needed better engagement at the programme level. Collectively, the various OPN programmes needed to focus more on impact, engage more with the UNEP Secretariat and bring new stakeholders on board with adequate resourcing.

There were challenges associated with COVID-19, particularly the move to virtual meetings, which served to maintain the programme's momentum.

Some of our MACs had this to say about the engagement of the SBC programme:

Seven years of SBC is a drop in the ocean. (CSO1)

Maybe what could be stronger is also the involvement of the private sector, largely so the companies, the stakeholders in the value chain that ultimately have to implement these principles. I definitely think with the work you have put out, you have developed a lot of understanding. You have developed a lot of best practice reports and really helped stakeholders get a grip of the issue. And I think that that's a strong element of the programme. (IND4)

Leadership

Having the Ministry of the Environment of Finland as the Lead of the SBC programme was a natural progression when OPN came into existence, as Finland was already engaged in this sector via previous work and demonstrated a track record in sustainable buildings. The Marrakesh Task Force[1] may be considered a precursor to the SBC programme as Finland played a leading role in the task force and supported the momentum leading to the programme. Finland, as Lead of the programme, was also involved in defining its scope, at least at the start, as well as the links with SCP overall, but these were further honed when the work plans were rolled out across the seven years of the programme.

Limited engagement of the OPN Secretariat in the work of the SBC programme was considered a problem. There were also communication problems, leading to a lack of trust, and some of the SBC CD members felt there was no spirit of collaboration. This undermined commitments and got in the way of a collective impact across the OPN programmes. Also, the role of the 10YFP as a framework remained unclear to many beyond it. Even though the name change occurred at some point, knowledge of the 10YFP was with a select group. Respondents highlighted that the programme had done some amazing work with reports, the development of tools, joint conferences and seminars (such as sustainable building conferences) and representation at major conferences – such as WUF, WCEF and COP – that was not widely shared, even within UNEP itself. There was no shortage of ideas on what the programmes needed to focus on, but the organisational structure of the OPN did not support implementation of these ideas. Each programme operated in silos with very little engagement between each other and with the National Focal Points as well as the 10YFP Board.

Many participants agreed that fragmentation existed in the OPN. Some programmes received considerable earmarked trust funds from individual governments to initiate implementation projects. At the same time, the means of the OPN Secretariat to support the programmes in fundraising were insufficient. It is important to raise awareness, but also to follow through with action and link into existing networks. With so many networks in place, it was essential to ensure there were no duplications and to use the existing networks as conduits to deliver action. This required time for mapping, and associated resources to ensure that mapping was correctly undertaken.

A focus on the rapid city-building regions of the world from the commencement of the SBC programme would have probably accelerated its impact. A focus on mandatory building codes and regulations, and contextualised responses that consider climate and culture, such as passive design, use of local materials and local technologies, plus utilising economies of scale for accelerating transitions, may have brought quicker returns and rewards. A clear spotlight on SMEs and the right policy coverage and context to support the people on the ground may have turned the attention more to grounded results.

Different approaches to circularity needed to be considered in some detail for the Global North and Global South. In addition, it was essential to consider not just the large industry supply chain but also the SMEs and the role of labour markets. Tied to labour are small-scale considerations: jobs and livelihoods for low-skilled or unskilled workforces. Discussions on issues such as these took place at the MACs but were not elevated or uplifted to support catalytic change. Jobs and the economy needed to be considered differently in the context of the Global South, along with gender and business models.

Those respondents that worked with industry and government felt that there should have been a focus on training, which again is very critical for the Global South. Training also needs to be undertaken on a rolling basis as people may move once trained; it also needs to be undertaken in a way that can be embedded into the workplace. Working directly with educational institutions, supply chains and at the policy levels would have permeated through the industry and elevated impact. More South–South exchanges would have been beneficial. Some of the tools developed could have been used for building countries' capacity for teaching students and professional development.

Material circularity has now permeated into the thinking in some countries in parts of the Global South such as Africa, as was reported by a MAC member.

In particular, some MACs were able to influence governments regarding the circularity agenda as they worked closely with some governments in various countries. But where there was no funding to take it further, it ended up becoming lost. The exception to this was the project funded through BMZ Germany, where the proposal was submitted by the SBC programme. Funding was awarded towards the end of the programme, managed and hosted by UNEP.

Some of the interview respondents underscored the points about leadership with the following quotes:

> But if you go outside this [SBC] family, there's a lot of governments and institutions who just don't know anything about it … I would say we are more or less post offices, taking or receiving all kinds of notes every now and then … not very often, but every now and then from the Secretariat and then depending on the person [and] how much time she or he has for [directing messages]. (GOV3)

> I think that of course all of us could look in the mirror also and ask whether we should do something better. (GOV1)

Funding

Funding is a challenge across any programme, but funds are critical to get changes on the ground. Generally, it was felt that the programmes needed to take too much initiative with regard to funding. At the same time, fundraising requires specific skills.

Based on 2019 data, it was noted that SCP does not have the same resource mobilisation as the other SDGs. Some SDGs, such as SDG 2 on zero hunger and SDG 3 on good health and well-being, received significant funds (SDG Financing Lab 2023). SDG 12 received the second lowest amount amongst all the SDGs, the lowest being SDG 14 on life below water. Although the focus of the SBC programme on developing countries was good, it did not naturally attract donors to ensure there were enough funds available for undertaking projects.

Despite the lack of funds, all respondents agreed that the programme achieved a lot. With limited funds, the achievements were excellent; outputs and outcomes demonstrated high levels of achievements. The interviewees felt that SBC should continue to focus on these into the future. More funding would have also assisted in programme promotion such as the state-of-play reports that were produced. Alignment of funding is also critical to ensure continuity. If supported by governments it is easier for the private sector to engage, as it is considered to be low-risk. So, the funding needed to be undertaken at both the national government level and private sector level.

The SBC programme did not get the full picture of the funding of the 10YFP programmes, so it is difficult to know exactly the scope of the budget the Secretariat was working with. Early on, as some participants noted, promises were made on resource mobilisation, but they never eventuated, although the Secretariat did make some efforts. There were not enough funds for operating the SBC programme properly; thereby, it did not achieve the wider scale of impact anticipated. Trust Fund projects helped, but they remained as seed projects; there was no opportunity to replicate and upscale. The funding from major donor organisations such as the World Bank (WB) and International Finance Corporation (IFC) never included funding for SBC, despite the programme trying to receive funds, as explained in Chapter 4.

Fundraising also needed to be done jointly, as a collaborative process with input from the six OPN programmes. Not only the SBC programme Lead but some of the MACs were fully supportive of finding donors for financing projects. The Government of Finland committed some funding to the SBC programme. In fact, it was the biggest single international project that the Ministry of the Environment of Finland supported at that time. In the seven years (2015–2022) of the programme, the funding provided by Finland was limited to travel for MAC meetings for delegates from developing countries, travel to international events such as the WUF and seed funding support for the state-of-play reports (2019–2020) and the deep dives across Asia, Africa and Latin America the following year (2021).

The budget associated with the Trust Fund was a one-off; if there was a rolling budget, that would have assisted in keeping the projects going. Trust funds could have been upscaled and replicated but this did not happen as there were no funds for the same. Some projects did demonstrate success, such as the certifications for the CSC in the LAC region (see Chapter 4). Essentially, the funding context was not understood by the SBC programme and the wider OPN as there was no business development expert involved in the broader OPN programme.

Interview participants had these comments about the funding of the SBC programme:

> It did what it can do with the very, very limited resources it has. But we should not forget that it is part of a system, and it has a role to play within the system. And there comes the connection with the other topics, the other programmes. The connection with housing, the question with quality of life, lifestyle, and the question of the system as a whole. The SBC programme did add value to the overall system of sustainability. But if we want to reach impact, sustainable building activities should be accelerated, multiplied and scaled up. (IGO3)

> Lack of funding, and lack of link to national governments. Maybe those are the lacking instruments or lacking links. (GOV3)

In hindsight, one of the Co-Leads stated:

> If Finland had been able to finance or seconded a person in the Secretariat like some other programmes have been able to do then it would have been different because then we would have had a dedicated person for the SBC programme. With our funding ... we would have had also a say on his or her scope of work. (GOV2)

Regarding resources already invested, GOV2 added:

> So this is like almost 1,000,000 EUR over the seven years that the Ministry of the Environment of Finland has invested in the SBC programme. And, I think it has been worth every penny. We now have an opportunity to really do better where we want to and should have results and activities that we want to continue with and what we can do in a different way.

Governance

Perhaps the greatest challenge has been in the governance of the OPN across the board. The SBC CD felt at times that the operations of the Secretariat were not transparent; it was hard to know what the wider strategy of the Secretariat entailed. From the perspective of the SBC programme, the Secretariat did not focus on supporting the implementation of the programme and coordinating better the work across the OPN. It was felt that there was a lack of strategic thinking and consistency in the governance across all the six OPN programmes. There were suggestions from the Secretariat on how the programmes may operate, but eventually each of the programmes ended up with one that worked for it at an operational level. The SBC programme managed to deliver well despite the challenges, while at the same time realising there were missed opportunities that better-resourced and better-managed OPN coordination and governance could have provided for the whole OPN family.

Another big challenge was that there were several initiatives on sustainable buildings and construction: the SBC programme, the GlobalABC and earlier the SBCI. The biggest issue was the governance of the SBC programme and the GlobalABC within the same unit of UNEP. The GlobalABC managed to raise significant funds compared to the programme and that became the focus rather than SBC, even though the GlobalABC was set up after the SBC programme. This led to challenges in equal management and provision of services by UNEP to the initiatives. These challenges were discussed by the management of the Ministry of the Environment of Finland and UNEP, but only temporary solutions were found.

The GlobalABC's annual outputs included Global Status Reports (GSR). These were helpful to track the progress over time and raise the value-add of the GlobalABC's work. This is highlighted in the resourcing of the GlobalABC as the yearly reports were funded by the donor countries and resources (consultants) were employed to produce these. These annual status reports were eventually launched at international meetings such as the COPs. Toolkits and reports developed by the SBC programme and by the OPN had limited reach, and the perception was that they did not have the same coverage as the GlobalABC.

Another aspect that further led to confusion was the perceptions of the roles of the SBC and GlobalABC programmes. At first glance they appeared to be in conflict, but the aims of the two programmes were different. The GlobalABC had the same players as the SBC programme, which added to the perception that the two programmes were in conflict. The GlobalABC's focus was on mitigation, whereas the SBC programme focused on circularity, also including mitigation and adaptation aspects.

The focus of the Secretariat was on quantitative reporting, in terms of expectations of what each programme would deliver (number of network partners, number of projects, donors, etc.) rather than focusing on the implementation and the overall outcomes that the OPN would collectively deliver. This eventually also led to ambivalence by the MAC members in the SBC programme, especially when donor funds were not forthcoming. One partner noted that administratively the Secretariat was struggling to deal with contractual matters and there was a lot of delay from its end.

Furthermore, the Secretariat was beset with changing personnel, alongside the competing initiatives and lack of resources. In theory, parallel initiatives can work together, but this needs to be well managed. As people were volunteering their time to work across the OPN, it was critical that the Secretariat manage the strategic focus, and drive the coordination and implementation. The investment of time and effort to get alignment is needed. Even if the Network meets and there is a good exchange of ideas, it does not always mean that something concrete is going to come out of it till someone drives the process. Resources are needed to set up strategy and link to the opportunities. Teamwork through collaboration, cooperation and synergies are needed. COVID-19 in the midst of the seven-year cycle caused further disruptions.

Governance considerations were highlighted by the Co-Leads as follows:

> It is complicated when you're trying to manage two networks or initiatives that both deal with the issue of sustainability in the buildings and construction sector, even if they both come from a different angle. Even though the SBC programme joined the GlobalABC when it was created there was not enough coordination and synergies between both initiatives. (IGO1)

> We haven't been encouraged to work together so much during these seven years. Maybe there should have been more encouragement for that [in reference to collaboration with other OPN programmes]. (GOV1)

GOV2 eloquently described the engagement and governance aspects of the SBC programme:

> Some people didn't understand that this is a process, and the One Planet Network is a community of practitioners, not a community of diplomats.

A MAC member had this to say:

> I see a lot of jockeying for positions, SBC versus GlobalABC. This is, of course, not very effective because these initiatives are all competing for funding, visibility and also for topics. This may force one or the other initiative to pursue a niche issue just to make it stand out. All of this is probably borne out of the very positive recognition that buildings are very important and the question of how to best deal with the sector within different parts of the UN hierarchy. I personally think bringing it all together might be a very good thing. This will actually allow focusing on areas that others are not working on. For example, if the Global Alliance is focusing on climate change and adaptation and resilience, then maybe whatever future programme might emerge might not necessarily need to have this as their main focus but merely follow that work stream very closely and use the learnings from the GlobalABC because otherwise the same situation may arise again. (CSO5)

Policy frameworks

Participants agreed that eventually the seed funding provided through the Trust Fund projects and other initiatives of the SBC programme was expected to be scaled up at country and regional levels. To do so, a range of measures to support policies through national governments and engagement is needed. Different approaches for different types of countries need to be canvassed. While local authorities can undertake construction, they work with central governments and ensure that there is alignment between action and policies.

Collaboration is needed to scale up. Though case studies were developed, the business case was not developed or not developed enough. The mass scaling up to respond to housing as an urgent need across the world has yet to happen. This needs to be undertaken by taking a holistic approach. The role of procurement in the process of delivering housing and working with policy frameworks are critical.

Some respondents noted that policy frameworks need to be supported by metrics to ensure ongoing monitoring and evaluation. This also involves understanding the supply chain (as materials that are being used) and understanding the labour markets, which have local nuances. While not always hitting the policy lever, there are project learnings and best practices seen through the SBC programme. Materials, best practice and standardisation are some mechanisms that have been picked up in the wider impact arena.

The role of policy has been highlighted by a MAC member and a partner:

> I think the thing[s] I've most taken heart from the meetings I've attended have been what people are actually doing ... From hearing some of these best practice stories from municipalities. Also, continued policy analysis and briefings seem pretty good examples of projects, and demonstration projects and reports that I imagine are passed on to people. (CSO3)

> I think SBC can help in many ways ... sharing experience knowledge definitively. And creating an enabling environment at a global level. (UNI1)

Impact

To deliver on impact, siloed approaches cannot work; systemic changes are needed, coupled with long-term strategies. The use of UNEP as a core partner was continuously considerably under-resourced despite discussions and promises. Also, the internal competition between the various UN units and a lack of resources created a difficult work environment. Working on circularity is complex and needs grounded approaches. An area of urgent need identified by some of the respondents was to make building ratings mandatory. Furthermore, a focus on procurement, energy efficiency and resource efficiency, and collaboration across all stakeholders; and designing good legislative and policy platforms, supported by finance, industry and science, were all identified as being critical to support impact. Tools are needed to support impact, and grants and funding can accelerate impact.

Understanding and communicating case studies is critical, and engaging with others to raise awareness and provide concrete examples, and the SBC programme has been successful in creating a network of networks. Some participants remarked that the perception across the OPN is that it has brought results.

Simple information at early stages of the project sets up the business case to make things clear to the investor and others in the supply chain. Developing the financial planning early is needed so that those who are specialists in sustainability can still get involved in the delivery of the project. The private sector needs to take the lead in some countries as well as working closely with the national government to ensure that NDC targets are met.

One partner noted the importance of seeking different financial models to scale impacts by working with the private sector:

> I mean the whole point of this, these partnership type models are to share risks and share benefits at the end. And I don't see any reason why they wouldn't work on a private basis equally well. Maybe even better, because private partnerships: you have two sides that might be more flexible in contracting than in a public–private partnership. (CSO4)

Yet others expressed their views differently:

> I would say that in the end we've been relatively successful. (IGO1)

> After that, it's like you train people for the project and it's very difficult to follow up and see where things have ended up. (UNI1)

> I think what's really unique about the SBC programme is that it really honed in and focused on circularity and the built environment. I don't think there were any other reports out there that really were focusing specifically on circularity for the built environment part and for buildings; how we design and construct buildings. And there's so much yet to be done in that space. So, going beyond just thinking about the end-of-life aspect of the life cycle to thinking about during construction and things like design for disassembly, like design principles that can be incorporated in the early stages that have huge impact then on the potential for these enabling strategies to work. (UNI2)

SBC: The Next Phase

The SBC programme's legacy has now been tied in with the GlobalABC programme. The GlobalABC already has areas of focus associated with the programme and a new circularity stream comprising the legacy of the SBC programme has been set up. The programme's focus on circularity has been morphed into the GlobalABC. Discussions in 2022 with the Secretariat towards the end of the first cycle of the OPN programmes resulted in the development of the SBE: the Sustainable Built Environment initiative. The underpinning primary rationale behind this was to continue the momentum on the built environment as an important sector for decarbonisation, while also recognising the important role procurement plays in the buildings and construction process. The focus has since been changed to circularity and the name of the initiative to Circular Built Environment (CBE).

A working group with the same name, Circular Built Environment (CBE), has been established within the Materials Hub of the GlobalABC, supported by a partnership between the OPN and GlobalABC secretariats, hosted by UNEP. The CBE is leading two projects:

1. A Circularity Assessment Framework that aims to develop and pilot a standardised approach to assessing circularity in the buildings and construction sector at the national level. Grounded in the SDGs and internationally recognised indicators and metrics, the project also involves crafting tools and guidance, with a special focus on implementation in developing economies. Piloting activities are underway in Senegal and Bangladesh to inform and refine this framework; and
2. The flagship initiative Mainstreaming Circularity in the Construction Sector through Sustainable Public Procurement, which is actively explor-

ing the potential of SPP as a catalyst for advancing circularity in the construction sector. The initiative aims to (1) develop a national commitments framework and a multi-stakeholder partnership; (2) map core organisations and initiatives, key indicators and metrics, networks, best practice policies and guidance; and (3) develop a systemic approach to support local governments and countries to implement commitments in different situations. The Co-Leads of the new OPN/GlobalABC Circular Built Environment working group are part of the strategic advisory group of the SPP flagship initiative together with the Co-Leads of the OPN SPP programme.

Moving forward, financed by Finland's recent and ongoing commitment to the CBE initiative, the OPN Secretariat has recruited a consultant who coordinates between the OPN and GlobalABC secretariats, supporting the CBE working group, led by the Ministry of the Environment of Finland and RMIT University. At the time of writing, this resource is based in Iceland, and progress deliverables for CBE will be guided by UNEP and the OPN Secretariat. Finland has also committed to supporting the CBE programme (in a different capacity than it has been supported to date) and other arms of the Finland government have also reiterated their support. Notwithstanding this, there is a certain level of disappointment amongst the SBC community about the SBC programme becoming part of the GlobalABC as a subcomponent, as the GlobalABC and SBC were sitting parallel with each other. For the Government of Finland, this was the best solution to keep the SBC programme's legacy alive.

Openness and transparency are needed to move forwards, along with ways to improve communication and implementation. Cross-sectoral benefits and synergies need to be explored. At the country level, focusing on mandatory regulation, especially where it does not exist, is critical to lay strong foundations for SBC. A grassroots focus needs to be encouraged as it is easy to focus on local solutions within the local areas.

With the hindsight of past experience, the following points need to be considered:

- Clear coordination and leadership, and engagement within the members of the programme itself and other OPN programmes, need clear direction. The role of stakeholders and network partners needs to be clearly articulated.
- Appropriate funding and resources are needed for programme implementation and for projects.
- The SBC work already undertaken needs to be built upon further to support upscaling.
- Work in the Global South needs to be expanded; conditions to support circularity have yet to be harnessed in some countries. Locking in good

practices where the most impact can be gleaned needs to be mapped. Construction for the masses, not just the elite, is needed.

- Along with materials, a focus on ownership models, finance mechanisms and the creation of material marketplaces needs nurturing. Major companies are signing up to circularity and to the SDGs and this will filter sooner or later into the Global South.
- Marketing and slick web designs are important. Text is important, but many people do not read. Visual cues and other forms of communication are needed for them to understand, and to ensure communication is on point.
- Focus is required beyond the building, to include the wider precinct scale and more, procurement, people and communities. Systems thinking approaches need to be applied. The new programme should continue to work on policy and set up collaboration platforms so that housing and other types of buildings, construction methods and processes, and infrastructure are all working in tune across the supply chain. Resilience and adaptation alongside mitigation is essential. The role of the new programme involves not just the global perspective but also the local, through grassroots engagement supporting SMEs, as it is easier to change a small firm than a large organisation. SBE needs to consider sequential and simultaneous engagement with the supply chain.
- Mindsets need to change; hence education and the development of new models supporting the transition, including business models, are a priority.
- Setting up material passports to underscore the importance of material impacts underpins circular thinking.
- The role of academia in countries is critical to build capacities and bring in evidence bases. Likewise, the new generations of students and decision makers need to understand sustainability and circular economies.

The momentum of SBC needs to continue. This has been noted by MACs as follows:

Overall an exciting journey for SBC, but it should build further rather than close. (CSO1)

Really the kind of translation of this information to decision makers, and those decision makers are normally investors helping the industry to make this a reality, so that it's not just kind of a conceptual design subject, because I think there's been some feedback that even if those designs were suggested, ultimately the decision comes from investors, and if they say no, basically the whole thing can be completely destroyed by that one decision. So how do we translate all this amazing information so that investors can make better-informed decisions? It's similar, I would say, to benchmarking, for example, and measurement of impact. (IND2)

It's gone into practice; we rewrote a curriculum, we developed training materials, we looked at bylaws, we looked at quite nice case studies. They are quite big pilot projects. We had 10 very different house projects using very different technologies that we evaluated and published. We had quite a lot of interaction with small-scale entrepreneurs that manufacture things like timber window frames, compressed earth construction, different interlocking blocks used in walling, so I think probably Zambia has been one of the areas where we've had the most impact. There is some work in Sierra Leone. I think some of the kinds of sharing [of] some of the conferences and the kind of larger-scale shared information, that's being a way of improving awareness. (IND5)

I think my view is that the need hasn't gone away. There's a huge need for research, thinking, networks, guidance, training, all of those sorts of things here definitely in the whole of Africa. There's a huge need for it, so I think an organization that facilitates projects and networks and linking is very useful, and I've found it very useful. I think it's a good thing to have something that can do your main day-to-day work. But you know there's a way of working across continents and being exposed to different people and different things. So, yeah, I would say ... it's [a] valuable network and, ideally, it carries on. If it needs to change, hopefully it doesn't lose some of the good things. (IND5)

Linking SDGs, equality, nature, local markets, resilience ... all the various bits and pieces ... What we mean with nature-positive for the built environment and next year, based on our generic work on inequality, we also want to look at that in the built environment, so I think SBC, with its focus on the SDGs and broader circularity and resource efficiency ... I think the whole nature discussion would be interesting because it is obviously very closely linked. And to, maybe again, help us almost draw a better common approach to that and then ultimately inequality. (IND4)

CONCLUSION

The IPCC has shown the urgency, importance and value of various things, namely:

- engaging through capability strengthening
- whole-of-life thinking
- the role of embodied carbon and materials and similar things.

This chapter has reaffirmed that urgency and confirmed the significance and scale of accelerated action that is needed.

The SBC programme has started the process, but the vision alone is insufficient. Action on the ground is needed. Despite limited funds, the SBC programme has achieved valuable outputs and outcomes. However, these need to be nurtured further to achieve scaling up for lasting changes. A dedicated effort towards marketing and supporting research translation is critical in the future of the buildings and construction sector.

As the role of building materials is being recognised as being critical, the role of digitalisation and technology also needs to be considered. Tools need to be translated to suit local conditions and into local languages to increase uptake. Benchmarking for circularity needs to continue, and some foundational principles of what constitutes circularity should be clarified. It is essential to get investors involved from the start, and to translate knowledge to support decision making, and greater alignment between governments, collaboration and the involvement of both public and private organisations are needed. Capacity building across the full value chain, including clients, needs to be urgently addressed.

NOTE

1. The Marrakesh Task Force stems from the Marrakesh Process launched in 2003 as a global effort to promote progress on the implementation of more sustainable consumption and production patterns. See https://www.un.org/esa/sustdev/marrakech/gpaper3chap3.pdf.

REFERENCE

SDG Financing Lab (2023) OECD Launches Tool to Analyze SDG Finance. SDG Financing Lab. https://sdg.iisd.org/news/oecd-launches-tool-to-analyze-sdg-finance/#:~:text=The%20 OCED%20developed%20the%20SDG,development%20financing%20across%20the%20 SDGs

APPENDIX: QUESTIONS SEEKING FEEDBACK ON THE SBC PROGRAMME ACROSS THE DIFFERENT STAKEHOLDER GROUPS (INTERVIEW QUESTIONS)

Lead and Co-Lead

1. How do you see the role of a) circularity and b) embodied carbon in sustainable building globally?
2. What is needed locally and globally from an embodied carbon and circularity perspective (resource efficiency, etc.) to support the future in building and construction?
3. What is needed to promote embodied carbon and resource use (on government level, industry level)?
4. How can a global programme like SBC help?
5. What worked well in the programme?
6. What were the challenges faced by the SBC programme?
7. What did not work well in the programme (focus on governance considerations)?
8. Does the OPN and in particular the SBC programme respond to the SDGs? Elaborate on how you think this is the case.
9. Do you have any comment to make on resource mobilisation?
10. Was the SBC programme well resourced?
11. How did the programme contribute to the overall SDG targets?
12. Were the SBC programme projects successful in contributing toward achieving the OPN/SDG targets?
13. What do you think were some of the financial challenges faced by the SBC programme? What impacts do you think this resulted in?
14. How do you think the relationship was with: MACs, NFPs, Board, Programme Partners, other OPN programmes?

MACs

1. How do you see the role of a) circularity and b) embodied carbon in sustainable building globally?
2. What is needed locally and globally from an embodied carbon and circularity perspective (resource efficiency, etc.) to support the future in building and construction?
3. What is needed to promote circularity (on government level)?
4. How can a global programme like SBC help?
5. What worked well in the SBC programme?
6. What are the challenges and barriers faced?

7. What contributions did you/your organisation make to the SBC programme?
8. Does the OPN and in particular the SBC programme respond to the SDGs? Elaborate on how you think this is the case.
9. What, if any, policy outputs did your organisation produce? Were any undertaken in association with SBC (what was the process and outcome/s)?
10. What changes, if any, do you observe in practice to implement sustainability in the sector?
11. Have your practices impacted: industry, academia, government? If so, how?
12. Has engagement with the SBC programme helped the country as a whole? If so, how?
13. How do you think the relationship was with: MACs, NFPs, Board, Programme Partners, other OPN programmes?
14. Was the engagement across the MACs enough? (Comment: relationships developed, time between meetings, contributing to projects, etc.)
15. Are there any particular themes that you would like SBC to engage with?
16. Are there any particular types of outputs/outcomes you would like the programme to focus on/engage with?

OPN SBC Programme Partners

1. How do you see the role of a) circularity and b) embodied carbon in sustainable building globally?
2. What is needed locally and globally from an embodied carbon and circularity perspective (resource efficiency, etc.) to support the future in building and construction?
3. What is needed to promote circularity (on government level)?
4. How can a global programme like SBC help?
5. Does the OPN and in particular the SBC programme respond to the SDGs? Elaborate on how you think this is the case.
6. Do you see the SBC programme as a high impact sector environmentally? Give reasons for your response.
7. Why did you/your organisation join the SBC programme as a partner?
8. What value did you gain out of engaging with the SBC programme?
9. What expectations did you have prior to joining the SBC programme and were they met?
10. Will you continue your engagement with the SBC programme in the short (3 years) or long term (beyond 3 years)?
11. Are there any particular themes that you would like SBC to engage with?

12. Are there any particular types of outputs/outcomes you would like the programme to focus on/engage with?

Secretariat

1. How do you see the role of a) circularity and b) embodied carbon in sustainable building globally?
2. What is needed locally and globally from an embodied carbon and circularity perspective (resource efficiency, etc.) to support the future in building and construction?
3. What is needed to promote circularity (at government level and industry level)?
4. How can a global programme like SBC help?
5. From your perspective, what were some of the challenges faced by the SBC programme?
6. In your opinion, did the SBC programme achieve its objectives?
7. In your view, were the SBC projects successful in achieving the SDG targets?
8. Does the OPN and in particular the SBC programme respond to the SDGs? Elaborate on how you think this is the case.
9. How did the programme achieve and implement the SDG targets and strategies respectively?
10. How did the programme and its projects contribute toward the SDGs and overall goals of sustainability (climate mitigation and adaptation) goals?
11. What were your expectations of the SBC programme?
12. In your view, what value has the SBC programme added to the overall OPN?
13. Are there any particular themes that you would like SBC to engage with?
14. Are there any particular types of outputs/outcomes you would like the programme to focus on/engage with?

National Focal Points

1. How do you see the role of a) circularity and b) embodied carbon in sustainable building globally?
2. What is needed to promote circularity (on government level if govt or choose industry)?
3. How a global programme like SBC could help?
4. What is needed locally and globally from an embodied carbon and circularity perspective (resource efficiency, etc.) to support the future in building and construction?

5. What are some of the challenges faced by the NFPs in working with the OPN programme?
6. What are some of the challenges faced by the NFPs in working with the SBC programme?
7. What, if any were some of the challenges faced by the NFPs in engaging with the SBC programme?
8. What role, if any, did you play with the SBC programme?

Other OPN Programmes

1. How do you see the role of a) circularity and b) embodied carbon in sustainable building globally?
2. What is needed locally and globally from an embodied carbon and circularity perspective (resource efficiency, etc.) to support the future in building and construction?
3. What is needed to promote circular practices (on government level)?
4. How could a global programme like SBC help?
5. What worked well in terms of cross collaboration programmes?
6. What did not work well in terms of cross collaboration?
7. What are some of the cross-programme collaboration challenges faced?
8. In which area/s do you see if any points of intersection between your programme and SBC?
9. Are there any particular types of outputs/outcomes you would like the programme to focus on/engage with?

7. Conclusion

Usha Iyer-Raniga

The SBC programme was aimed at improving knowledge of sustainable construction and supported this work so as to mainstream sustainable building solutions. Through the programme, all major sustainable construction activities were brought together under the same umbrella. The work involved sharing good practices, launching implementation projects, creating cooperation networks and committing actors around the world to sustainable construction. This final chapter presents the conclusions of the work undertaken by the programme over the period 2015–2022.

Chapter 1 discussed the importance of the SBC programme and the role of buildings and construction in alleviating our carbon emissions. This is a particularly critical consideration in the new growth regions of Asia, Africa and Latin America, which are facing rapid city-building activities and increased urbanisation. Cities contribute to nearly three-quarters of our emissions; a significant proportion of this comes from buildings. The buildings sector can contribute to reduced GHG emissions through increased efficiencies in operation, but also improvement in the materials used and their embodied emissions through careful selection of materials. Whether it is in deep renovations in the Global North or new approaches to planning, designing, constructing and operating buildings in the Global South, the critical role of circular approaches in buildings and construction should not be undermined.

Chapter 2 presented a brief history of the programme and the 12 guiding principles that it was founded upon. From the very start, the programme's objective was to develop work streams that focused on sharing policies and other such enablers, supporting and promoting sustainable housing, enhancing sustainability in the supply chain, and reducing climate impacts by strengthening resilience in the buildings and construction sector. Promoting awareness, dissemination of the work undertaken, supporting a network of global and regional groups on lessons learned, and promoting interdisciplinary engagement across the various disciplines comprising the built environment, while also focusing on demonstration projects and case studies, were the themes that underpinned the programme.

The programme's structure was robust, with a government as Lead and research and intergovernmental organisations as Co-Leads. Over the seven years of the programme, four work plans guided its outputs and overall outcomes. The programme shifted its emphasis from buildings to the built environment in 2019, when the focus was on affordable and resilient built environments and responsibly sourced materials in a circular built environment. Through this, the programme sought to make the case for SBC in countries; provide training, and technical and policy guidance; and facilitate the creation of partnerships for implementation. The SBC programme set up a theory-of-change framework to deliver on these outputs and drive its sustainable and circular outcomes.

Chapter 3 examined the role of the value chain in the building and construction process. The value chain and the actors in the construction life cycle were defined to understand how they can add value to TBL sustainability. The standard whole-of-life approach includes pre-construction, manufacturing and product stages, construction, use and end of life. However, to be cognisant of reuse of materials post use or in their second life, the standard system boundaries are not considered. This is the case of the classic scope 3 approach in calculating GHG, where indirect emissions arising from downstream activities are not known. Since scope 3 is opaque and complex, the SBC programme's approach was to encourage circularity in built environments through the life cycle of buildings in six phases: manufacturing, design, construction, operation and use, renovation, deconstruction and end of life. Circular principles may be applied across any of the stages to prolong the life of buildings or use building materials in their second (or more) life, or any of the other approaches as described in the chapter.

While the work undertaken by the OPN Secretariat in consultation with the IRP showed the stages of the construction value chain as a linear process and noted both processes and actors as part of the value chain, the SBC programme clearly distinguished the life cycle process, the actors and the value created by the different stages in a circular process. Rather than just environmental and economic considerations regarding sustainability, the SBC programme also focused on the role and the importance of social value, such as job creation, a critical value-add in the vast informal sector associated with construction in the Asian, African and Latin American regions.

Furthermore, the role of operations or the use phase is shifting to emphasise the importance of the construction phase of buildings because of the proportional impact that embodied energy and materials have in the construction phase of the building life cycle. Therefore, the focus cannot simply rest on energy efficiency alone; the full life cycles of buildings need to be considered now and in the immediate future.

In Chapter 4, five SBC 10YFP Trust Fund projects covered widely the topics of sustainable buildings and construction based on a general call at the end of 2015. Each of the projects were situated in the Global South. The projects focused on policy (Colombia), tool development for housing (Burkina Faso, Kenya and Nepal), building materials (India), planning (Egypt) and green concrete production (Colombia, Costa Rica and Panama). Overall, the programme contributed to these projects with a total budget of 668,000 USD. While each of these projects had immediate outputs, they also needed to be scaled up, to optimise the impact of the work undertaken. This required the injection of funds and needed dedicated time and effort for business development, which was challenging for the SBC programme.

In 2018, the MPTF was set up by the Secretariat, representing a partnership between UNEP, FAO, UNWTO, UN-Habitat, UNOPS and UNDP. Through this fund, the RE-Think Buildings toolkit comprising a three-step tool to help governments to transform their buildings and construction sector to a resource-efficient, low-carbon and climate-resilient one was developed by UNEP with UN-Habitat and UNOPS through the SBC umbrella with advice from SBC partners.

In the absence of other internationally accepted frameworks, the SBC programme decided to use the SDGs as the vehicle for mapping and tracking circularity in the built environment. Chapter 5 explored the SDG mapping against the circularity focus of the SBC programme using a global survey. Since the survey was skewed towards the Global North, workshops were undertaken in the Global South to ensure that the findings were representative. Based on the surveys and the expert workshops, four SDGs were identified as crucial for circular built environments. These are SDG 9 on resilient infrastructure, promoting inclusive and sustainable industrialisation and fostering innovation; SDG 11 on making cities and human settlements inclusive, safe, resilient and sustainable; SDG 12 on ensuring consumption and production patterns (and the foundation of the OPN); and SDG 13 on taking urgent action to combat climate change and its impacts.

Some, but not all, of the indicators under each of the goals were identified as being associated with circularity and the built environment. A core set of indicators alongside a secondary set of indicators were identified as being critical for circular built environments. In addition, with respect to the Global South, other goals – such as SDG 6 on ensuring the availability and sustainable management of water and sanitation for all; SDG 7 on ensuring access to affordable, reliable and sustainable energy for all; and SDG 8 on promoting sustained, inclusive and sustainable economic growth with productive employment and decent work for all – were identified. This is logical since the Global South is still struggling with access to basic resources and engaging in employment.

Chapter 6 presented the feedback from various stakeholders involved as part of the programme and explained the development of the next phase of SBC under the OPN. It has already been flagged that the SBC programme would be partnered formally with the SPP programme to form the new SBE programme.

The mandate of the 10YFP came to an end in December 2022. However, the UN General Assembly has extended the mandate of the 10YFP to 2030. For the OPN, it was necessary to plan for post 2022 and to discuss options for continued multilateral cooperation on SCP, and a legacy of the 10YFP and OPN to progress on the implementation of the 2030 Agenda and SDG 12 in particular. A renewed mandate is an opportunity that requires strong leadership, stakeholder engagement and commitment.

Finland confirmed its commitment to the leadership of the SBC programme until 2022 and willingness to build an ambitious vision for multilateral cooperation on SCP after 2022 together with the Friends of SCP. The Friends of SCP is a voluntary and informal coalition aimed at supporting the 10YFP Board and gathering the Member States that have been leading the 10YFP and its thematic programmes and continue to be willing to take an active role in building a vision on SCP after 2022.

Finland has reminded us that thematic programmes, like the SBC programme, have worked tirelessly over the last seven years and built precious worldwide communities of experts in their fields, be they construction, tourism, procurement or food systems. The relationship between the framework programme and thematic programmes has evolved over the years. Therefore, it was important to find a feasible balance between policy and practice in the post-2022 planning and implementation.

The ultimate intention is to take the post-2022 vision and decision making to the highest possible level and to accelerate actions for SCP. OPN collaboration opportunities exist in SPP through circular procurement; STP by supporting new and improved building technology and the use of local materials in tourism resorts; changing user behaviour and good life goals in buildings and neighbourhoods as part of the SLE programme; increasing knowledge, awareness and biodiversity for product information aligned with CI-SCP; and finally through greening opportunities arising from nature-based solutions to support regeneration and exploring opportunities for local food production in the planning and design of buildings and neighbourhoods where possible through the SFS programme.

The new SBE programme can build afresh from the learnings of the past and look towards a future that can continue to build the momentum that the SBC programme has set in motion. Despite the challenges associated with funding, the feedback received from the stakeholders of the SBC programme was that it was ahead of its time and delivered an impact. The modest impacts arising from the SBC programme can now be nurtured to be scaled up, replicated and mainstreamed, particularly in the Global South. Furthermore, the work seeded across circularity and attendant mapping against the SDGs are a good starting point for focusing on certain types of infrastructure needs, such as housing, in the countries of the Global South. Going forwards, the new CBE group working under the umbrella of the GlobalABC can progress the momentum of the SBC programme. The 10YFP Board supports this approach, as determined by its decision to continue the work of the SBC programme in a new partnership with the GlobalABC, through the adoption of the global strategy for SCP (2023–2030). The strategy also strengthens the role of the Secretariat in the coordination of all the existing OPN programmes.

As the effects of climate change are beginning to be felt across the world, it is clear there is no time to waste. The renewed programme needs to both continue the work and, through targeted dissemination and training, support the actors engaged in the praxis of CE in the buildings and construction industry. The new partnership underpinning the SBE programme, the alignment with the GlobalABC, and the financial support provided by the Government of Finland in 2022 to the programme set the tone for a new era of engagement and partnerships for taking the legacy of the SBC programme forward.

Index